BASIC REACTION KINETICS AND MECHANISMS

BASIC
REACTION KINETICS
AND
MECHANISMS

H. E. AVERY

Formerly Principal Lecturer in Chemistry
Lanchester Polytechnic, Coventry

M

First published 1974
Reprinted 1976, 1977

Published by

THE MACMILLAN PRESS LTD

London and Basingstoke
*Associated companies in Delhi Dublin Hong Kong Johannesburg Lagos
Melbourne New York Singapore and Tokyo*

ISBN 0 333 12696 3 (hard cover)
    0 333 15381 2 (paper cover)

*Printed in Great Britain at The Spottiswoode Ballantyne Press
by William Clowes and Sons Limited, London, Colchester and Beccles*

CONTENTS

PREFACE

This book is based on a course of lectures given at Liverpool Polytechnic to both full-time and part-time chemistry students. In the experience of the author, many students find final-year kinetics difficult unless they have thoroughly grasped the basic principles of the subject in the first year of study. This book requires no previous knowledge of kinetics, but gives a more detailed account than that found in general physical chemistry text-books. The purpose of this book is to give the reader a sound understanding of the fundamentals of the subject without dealing with the derivation of rate equations for typical complex reactions. The student is, however, introduced to a number of topics that will be dealt with more comprehensively in the final year of an Honours Chemistry course. With this in mind, the book should prove suitable for chemistry students in all years of Honours courses. It is particularly applicable to the first and second years of B.Sc. Honours Degrees, and for B.Sc. Ordinary Degree, Higher National Certificate, Higher National Diploma and Grad.R.I.C. Part 1 Chemistry courses. The book should also prove adequate for all the kinetics covered on B.Sc. combined science, biology, pharmacy and biochemistry courses.

The first four chapters cover the basic kinetic laws, the factors that control reaction rates and the classical methods used to measure reaction rates. These chapters cover the introduction to most kinetics courses. The basic theories of reaction rates are covered in chapters 5, 6 and 8, and a number of topics are introduced in a simple way. In chapter 7, the subject of atomic and free-radical reactions is treated comprehensively, since the author has found that the study of chain reactions has proved interesting and stimulating to many students; this area of kinetics has of course been a fruitful field of research for a number of years. The book concludes with a study of catalysed reactions, photochemical reactions and the development of new techniques for the study of fast reactions.

A number of worked examples are given throughout the book to illustrate the methods and relationships outlined. The reader is advised to test his knowledge of the subject on the kinetics problems that appear at the end of most chapters. For the diligent student, a list of key references, review articles and other textbooks for further reading are suggested at the end of each chapter.

Throughout the text, and for all the worked examples or test problems, the single system of SI units has been used. A selection of the units encountered most frequently in the book and a brief guide to the modern method of expression for physico-chemical quantities is given in an appendix.

The author wishes to express his appreciation to the students to whom he has taught kinetics at Liverpool Polytechnic and latterly at Lanchester Polytechnic, Coventry for their interest and keenness in the subject. I would also like to thank my wife for her continued help in preparing the manuscript and checking the text.

Some of the problems have been taken from past examination papers, and in this respect, I wish to thank the Universities of Brunel, Edinburgh, Hull, Lancaster, Liverpool, Manchester, Salford and Southampton for permission to publish.

H. E. Avery

1 INTRODUCTION

1.1 Kinetics and thermodynamics

The chemist is concerned with the laws of chemical interaction. The theories that have been expounded to explain such interactions are based largely on experimental results. The approach has mainly been by thermodynamic or kinetic methods. In thermodynamics, conclusions are reached on the basis of the changes in energy and entropy that accompany a change in a system. From a value of the free-energy change of a reaction and hence its equilibrium constant, it is possible to predict the direction in which a chemical change will take place. Thermodynamics cannot, however, give any information about the rate at which a change takes place or the mechanism by which the reactants are converted to products.

In most practical situations, as much information as is possible is obtained from both thermodynamic and kinetic measurements. For example, the Haber process for the manufacture of ammonia from nitrogen and hydrogen is represented by the equation

$$N_2 + 3H_2 \rightarrow 2NH_3 \qquad\qquad \Delta H^{\ominus}_{298\,K} = -92.4 \text{ kJ mol}^{-1}$$

Since the reaction is exothermic, le Chatelier's principle predicts that the production of ammonia is favoured by high pressures and low temperatures. On the other hand, the rate of production of ammonia at 200°C is so slow that as an industrial process it would not be economical. In the Haber process, therefore, the equilibrium is pushed in favour of the ammonia by use of high pressures, while a compromise temperature of 450°C and the presence of a catalyst speed up the rate of attainment of equilibrium. In this way the thermodynamic and the kinetic factors are utilised to specify the optimum conditions.

Similarly, in order to establish a reaction mechanism, it is helpful to consider all the thermodynamic and kinetic rate data that is available.

1.2 Introduction to kinetics

1.2.1 Stoichiometry

It is conventional to write down a chemical reaction in the form of its stoichiometric equation. This gives the simplest ratio of the number of molecules of reactants to the number of molecules of products. It is therefore a quantitative relationship between the reactants and the products. But it cannot be assumed that the stoichiometric equation necessarily represents the mechanism of the molecular process between the reactants. For example, the stoichiometric equation for the production of ammonia by the Haber process is

$$N_2 + 3H_2 \rightarrow 2NH_3$$

but this does not imply that three molecules of hydrogen and one molecule of nitrogen collide simultaneously to give two molecules of ammonia. The reaction

$$2KMnO_4 + 16HCl \rightarrow 2KCl + 2MnCl_2 + 8H_2O + 5Cl_2$$

tells us very little about the mechanism of the reaction, but this change can be represented by the stoichiometric equation since it gives the quantitative relationship between reactants and products.

In many reactions the stoichiometric equation suggests that the reaction is much simpler than it is in reality. For example, the thermal decomposition of nitrous oxide

$$2N_2O \rightarrow 2N_2 + O_2$$

occurs in two steps, the first involving the decomposition of nitrous oxide into an oxygen atom and nitrogen

$$N_2O \rightarrow O: + N_2$$

followed by the reaction of the oxygen atom with nitrous oxide to give one molecule of nitrogen and one of oxygen

$$O: + N_2O \rightarrow N_2 + O_2$$

This is a simple case in which the sum of the two individual or elementary processes gives the stoichiometric equation. Many other processes are much more complex and the algebraic sum of the elementary processes is so complicated as not to give the stoichiometric equation.

The thermal decomposition of acetaldehyde can be expressed as

$$CH_3CHO \rightarrow CH_4 + CO$$

But each acetaldehyde molecule does not break down in a single step to give one molecule of methane and one molecule of carbon monoxide. Kinetic results are consistent with a mechanism which proposes that the acetaldehyde molecule decomposes first into a methyl radical and a formyl radical. The products are formed by subsequent reactions between these radicals, acetyl radicals and acetaldehyde itself. The overall mechanism in its simplest form is

$$CH_3CHO \rightarrow CH_3\cdot + CHO\cdot$$
$$CH_3\cdot + CH_3CHO \rightarrow CH_4 + CH_3CO\cdot$$
$$CH_3CO\cdot \rightarrow CH_3\cdot + CO$$
$$CH_3\cdot + CH_3\cdot \rightarrow C_2H_6$$

The stoichiometric equation for the decomposition of dinitrogen pentoxide is

$$2N_2O_5 \rightarrow 4NO_2 + O_2$$

This is also a much more complex process than indicated by this equation and was shown by Ogg to proceed via the following mechanism

$$N_2O_5 \rightarrow NO_2 + NO_3\cdot \qquad (1)\,(fast)$$
$$NO_2 + NO_3\cdot \rightarrow NO_2 + O_2 + NO \qquad (2)\,(slow)$$
$$NO + NO_3\cdot \rightarrow 2NO_2 \qquad (3)\,(fast)$$

Kinetic studies showed that step (2) was the slowest stage of the reaction, so that the overall rate depends on the rate of this step and is therefore said to be the *rate-determining step*.

1.2.2   Molecularity
The *molecularity* of a chemical reaction is defined as the number of molecules of reactant participating in a simple reaction consisting of a single elementary step. Most elementary reactions have a molecularity of one or two, although some reactions involving three molecules colliding simultaneously have a molecularity of three, and in very rare cases in solution, the molecularity is four.

1.2.3   Unimolecular reactions
A unimolecular reaction involves a single reactant molecule, and is either an isomerisation

$$A \rightarrow B$$

or a decomposition

$$A \rightarrow B + C$$

Some examples of unimolecular reactions are

$\rightarrow CH_3CH=CH_2$

$$CH_3NC \rightarrow CH_3CN$$

$$C_2H_6 \rightarrow 2CH_3\cdot$$

$\rightarrow 2C_2H_4$

$$C_2H_5\cdot \rightarrow C_2H_4 + H\cdot$$

1.2.4   Bimolecular reactions

A bimolecular reaction is one in which two like or unlike reactant molecules combine to give a single product or a number of product molecules. They are either association reactions (the reverse of a decomposition reaction)

$$A + B \rightarrow AB$$

$$2A \rightarrow A_2$$

or exchange reactions

$$A + B \rightarrow C + D$$

$$2A \rightarrow C + D$$

Some examples of bimolecular reactions are

$$CH_3\cdot + C_2H_5\cdot \rightarrow C_3H_8$$

$$CH_3\cdot + CH_3\cdot \rightarrow C_2H_6$$

$$C_2H_4 + HI \rightarrow C_2H_5I$$

$$H\cdot + H_2 \rightarrow H_2 + H\cdot$$

$$O_3 + NO \rightarrow O_2 + NO_2$$

Sullivan[1] has shown that the frequently quoted 'classical bimolecular reaction

$$2HI \rightarrow H_2 + I_2$$

is a chain reaction at high temperatures (800 K) with the rate-determining step being termolecular.

1.2.5 Termolecular Reactions

Termolecular reactions are relatively rare since they involve the collision of three molecules simultaneously to give a product or products

$$A + B + C \rightarrow \text{products}$$

Some examples of termolecular reactions are

$$2NO + O_2 \rightarrow 2NO_2$$

$$2NO + Cl_2 \rightarrow 2NOCl$$

$$2I\cdot + H_2 \rightarrow 2HI$$

$$H\cdot + H\cdot + Ar \rightarrow H_2 + Ar$$

As can be seen from the examples given above, the term 'molecularity' is not confined to processes that involve stable molecules but is used when the reacting species are atoms, free radicals or ions. Therefore in the decomposition of acetaldehyde, the breakdown of the acetyl radical

$$CH_3CO\cdot \rightarrow CH_3\cdot + CO$$

is a unimolecular process, while the recombination of methyl radicals is a bimolecular process

$$CH_3\cdot + CH_3\cdot \rightarrow 2C_2H_6$$

If only effective in the presence of a third molecule (known as a third body) that takes up the excess energy, it is a termolecular reaction

$$CH_3\cdot + CH_3\cdot + M \rightarrow C_2H_6 + M$$

It is only appropriate to use the term molecularity for a process that takes place in a single or elementary step. The term therefore implies a theoretical understanding of the molecular dynamics of the reaction. Reactions in which a reactant molecule or molecules give a product or products in a single or elementary step are rare. If the reaction is a complex reaction, it is necessary to specify the molecularity of each individual step in the reaction.

1.3 Elucidation of reaction mechanisms

The ultimate task of a kineticist is to predict the rate of any reaction under a given set of experimental conditions. This is difficult to achieve in all but a few cases. At best, a mechanism is proposed, which is in qualitative and quantitative agreement with the known experimental kinetic measurements.

When a reaction mechanism is proposed for a certain reaction, it should be tested by the following criteria.

(i) Consistency with experimental results
It is easy to propose a mechanism for a reaction for which very little experimental information is available. In such cases it is difficult to prove or disprove the proposal. However, as more and more experimental data are obtained, it often becomes more and more difficult to find a mechanism that satisfies all the known results. It is only possible to be confident that a mechanism is correct when it is consistent with all the known rate data for that reaction.

(ii) Energetic feasibility
When a decomposition reaction occurs, it is the weakest bond in the molecule that breaks. Therefore in the decomposition of ditertiarybutyl peroxide it is the O—O bond that breaks initially giving two ditertiarybutoxyl radicals. In a mechanism in which atoms or free radicals are involved, a process which is exothermic or the least endothermic is most likely to be an important step in the reaction. In the photolysis of hydrogen iodide (see page 140), the possible propagation reactions are

$$H\cdot + HI \rightarrow H_2 + I\cdot \qquad\qquad (1) \; \Delta H = -134 \text{ kJ mol}^{-1}$$

and

$$I\cdot + HI \rightarrow I_2 + H \qquad\qquad (2) \; \Delta H = 146 \text{ kJ mol}^{-1}$$

For the endothermic reaction (2) to take place, at least 146 kJ of energy must be acquired by collisions between the iodine atoms and hydrogen iodide molecules. Reaction (2) is, therefore, likely to be slow compared to reaction (1).

If a mechanism involves the decomposition of an ethoxyl radical, the following decomposition routes are all possible

$$C_2H_5O\cdot \rightarrow C_2H_5\cdot + O\!: \qquad\qquad (1) \; \Delta H = 386 \text{ kJ mol}^{-1}$$

$$C_2H_5O\cdot \rightarrow CH_3CHO + H\cdot \qquad\qquad (2) \; \Delta H = 85 \text{ kJ mol}^{-1}$$

$$C_2H_5O\cdot \rightarrow CH_3\cdot + CH_2O \qquad\qquad (3) \; \Delta H = 51 \text{ kJ mol}^{-1}$$

$$C_2H_5O\cdot \rightarrow C_2H_4 + OH\cdot \qquad\qquad (4) \; \Delta H = 122 \text{ kJ mol}^{-1}$$

Again, the heats of reaction show that reaction (3) is likely to be the important process.

(iii) Principle of microscopic reversibility
This principle states that for an elementary reaction, the reverse reaction proceeds in the opposite direction by the same route. Consequently it is not possible to include in a reaction mechanism any step, which could not take

place if the reaction were reversed. For instance, in the thermal decomposition of ditertiarybutyl peroxide, it is not possible to postulate the initial step as

$$(CH_3)_3COOC(CH_3)_3 \rightarrow 6CH_3\cdot + 2CO$$

since the reverse step could not take place. Further, as all steps in a reaction mechanism are either unimolecular, bimolecular or termolecular, any proposed mechanism must not contain elementary steps that give more than three product species, for then the reverse step would not be possible.

(iv) Consistency with analogous reactions

It is reasonable to expect that if the mechanism proposed for the decomposition of acetaldehyde is well established, then the mechanism for the decomposition of other aldehydes would be similar. However, while it is in order to carry out similar experiments to prove this, it is dangerous to assume that a reaction mechanism is the same solely by analogy. Indeed there are numerous examples of reactions from the same series of chemical compounds that proceed via entirely different mechanisms, for example the hydrogen–halogen reactions. Analogy is therefore a useful guideline, but is not a substitute for experiment.

It can be appreciated that the more the rates of elementary reactions are studied, the greater will be the measure of confidence in the correctness of a proposed reaction mechanism. To obtain such rate data, modern techniques are used to determine the rates of very fast reactions and to measure very low concentrations of transient reactive species formed in reaction systems. A number of examples are given in later chapters of proposed reaction mechanisms based on kinetic data obtained by rate experiments. It is first necessary to establish the simple kinetic laws and a theory of reaction rates before proceeding to a study of more complex chemical reactions.

Further reading

References
1. J. H. Sullivan. *J. chem. Phys.*, **30** (1959), 1292; *J. chem. Phys.*, **36** (1962), 1925.

Books
E. L. King. *How Chemical Reactions Occur*, Benjamin, New York (1963).
R. A. Jackson. *Mechanism, An Introduction to the Study of Organic Reactions*, Clarendon Press, Oxford (1972).

Review
J. O. Edwards, E. F. Greene and J. Ross. From stoichiometry and rate law to mechanism. *J. chem. Ed.*, **45** (1968), 381.

2 ELEMENTARY RATE LAWS

2.1 Rate equation

Consider a chemical reaction in which a reactant A decomposes to give two products, B and C

$$A \rightarrow B + C$$

During the course of the reaction, the concentration of A decreases while the concentration of B and C increases. A typical concentration–time graph for A is shown in figure 2.1.

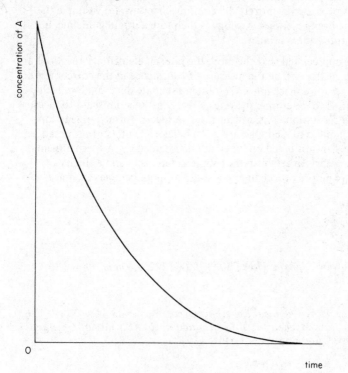

Figure 2.1 *Typical concentration–time curve*

Any rate is given by the change in a measureable quantity with time, and the *rate of a chemical reaction* is expressed in terms of the change in concentration of reactant in a given time. The rate of this reaction at any time t is given by the slope of the curve at that time, that is, is equal to the decrease in the concentration of A in a given time. Alternatively, the rate could equally be given by the increase in the concentration of B or C with time

$$\text{rate} = -\frac{d[A]}{dt} = \frac{d[B]}{dt} = \frac{d[C]}{dt}$$

The rate of a chemical reaction is therefore expressed as a rate of decomposition or disappearance of a reactant or the rate of formation of a product.

Figure 2.1 shows that the rate of the reaction changes during the course of the reaction. The rate, which is initially at a maximum, decreases as the reaction proceeds. It is found that the rate of a reaction depends on the concentration of the reactants, so that as the concentration of A in the above reaction decreases, the rate of reaction also decreases. Therefore,

$$\text{rate} \propto [A]^n$$

where n is a constant known as the *order* of the reaction. The relationship between the rate and concentration is called the *rate equation* and takes the form

$$-\frac{d[A]}{dt} = k_r[A]^n$$

where k_r is a constant for any reaction at one temperature and is called the *rate constant*. The rate equation states how the rate of an elementary step varies with the concentration of the reactants; the concentration of products is not involved in this expression.

2.1.1 Order of a reaction

If in the above reaction it is found by experiment that the rate is directly proportional to the concentration of A, the reaction is said to be first order, since

$$-\frac{d[A]}{dt} = k_r[A] \tag{2.1}$$

If the rate is found to depend on the square of the concentration of A, the reaction is said to be second order, since

$$-\frac{d[A]}{dt} = k_r[A]^2 \tag{2.2}$$

For a different process

$$A + B \rightarrow C + D$$

if the rate equation is found to be

$$-\frac{d[A]}{dt} = -\frac{d[B]}{dt} = k_r [A][B] \qquad (2.3)$$

the reaction is second order: first order with respect to A, and first order with respect to B.

In general for a reaction

$$A + B + C + \ldots \rightarrow products$$

$$rate = k_r [A]^{n_1} [B]^{n_2} [C]^{n_3} \ldots \qquad (2.4)$$

The order of the reaction is the sum of the exponents $n_1 + n_2 + n_3 + \ldots$; the order with respect to A is n_1, with respect to B is n_2 and with respect to C is n_3, etc.

2.1.2   Rate constant

The rate constant provides a useful measure of the rate of a chemical reaction at a specified temperature. It is important to realise that its units depend on the order of the reaction.

For example, the first-order rate equation is

$$-\frac{d[A]}{dt} = k_r [A]$$

Thus

$$\frac{concentration}{time} = k_r \, (concentration)$$

Therefore, for all first-order processes, the rate constant k_r has units of time^{-1}.

For a second-order reaction the rate equation is of the form

$$rate = k_r \, (concentration)^2$$

Therefore, a second-order rate constant has units of concentration^{-1} time^{-1}, for example $dm^3 \, mol^{-1} \, s^{-1}$.

In general, the rate constant for a nth-order reaction has units (concentration)$^{1-n}$ time^{-1}. From this it can be seen that typical units for a zero-order reaction are $mol \, dm^{-3} \, s^{-1}$, and for a third-order reaction are $dm^6 \, mol^{-2} \, s^{-1}$.

2.2 Determination of order of reaction and rate constant

The rate equations used so far in this chapter are all differential equations. If a concentration–time graph is drawn as in figure 2.1, the rate is measured directly from the slope of the graph. A tangent is drawn to the curve at different points and $-dc/dt$ is obtained. The initial slope of this graph gives the initial rate, and for a second-order process equation 2.4 becomes

$$(\text{rate})_{t=0} = k_r[A]_0[B]_0$$

where $[A]_0$ and $[B]_0$ are the initial concentrations of A and B respectively. An example of the use of this method to determine the rate constant is described in chapter 3.

Since the measurement of initial rates is not easy, it is preferable to integrate the rate equation. The integrated rate equation gives a relationship between the rate constant and the rate of chemical change for any reaction. The form of the equation depends on the order of the reaction. A summary of the different forms of the rate laws is given in table 2.1 on page 24.

2.3 First-order integrated rate equation

Consider the reaction

$$A \rightarrow \text{products}$$

Let a be the initial concentration of A and let x be the decrease in the concentration of A in time t. The concentration of A at time t is therefore $a - x$. The rate of reaction is given by

$$-\frac{d[A]}{dt} = -\frac{d(a - x)}{dt} = \frac{dx}{dt}$$

The differential rate equation, $-d[A]/dt = k_r[A]$, can be written therefore as

$$\frac{dx}{dt} = k_r(a - x)$$

or

$$\frac{dx}{a - x} = k_r \, dt \qquad (2.5)$$

Integration of equation 2.5 gives

$$-\ln(a - x) = k_r t + \text{constant}$$

Since at $t = 0$, $x = 0$, the constant is equal to $-\ln a$, so that substitution in equation 2.5 gives

$$k_r t = \ln\left(\frac{a}{a - x}\right)$$

or

$$k_r = \frac{1}{t}\ln\left(\frac{a}{a - x}\right) \tag{2.6}$$

Using logarithms to the base 10

$$k_r = \frac{2.303}{t}\log_{10}\left(\frac{a}{a - x}\right) \tag{2.7}$$

Equations 2.6 and 2.7 are obeyed by all first-order reactions.

2.3.1   Determination of first-order rate constants

(i) Subsitution method
The values of $a - x$ are determined experimentally by one of the methods described in chapter 3 at different times t throughout a kinetic experiment. These values are substituted in equation 2.7 and an average value of the rate constant is determined.

(ii) Graphical method
From equation 2.7 it can be seen that a plot of $\log_{10}(a/a - x)$ against t will be linear with slope equal to $k_r/2.303$ if the reaction is first order. Alternatively equation 2.7 can be rearranged to give

$$\log_{10}(a - x) = \log_{10} a - \frac{k_r t}{2.303} \tag{2.8}$$

A plot of $\log_{10}(a - x)$ against t will be linear with slope equal to $-k_r/2.303$. If the rate data obtained gives a linear plot the reaction is first order, and the rate constant is determined from the slope. A graphical determination of k_r is more satisfactory than method (i).

(iii) Fractional life method
For a first-order process, the time required for the concentration of reactant to decrease by a certain fraction of the initial concentration is independent of the initial concentration.
   Let $t_{0.5}$ be the time required for the initial concentration a to decrease

to half the initial concentration (0.5a). This is known as the *half-life* of the reaction. Therefore, for half-life conditions, equation 2.6 becomes

$$k_r = \frac{1}{t_{0.5}} \ln \frac{a}{0.5a}$$

$$= \frac{\ln 2}{t_{0.5}}$$

$$= \frac{0.693}{t_{0.5}} \qquad .99$$

or

$$t_{0.5} = \frac{0.693}{k_r} \qquad (2.9)$$

which is a constant for any particular reaction, and is independent of the initial concentration.

In general, the time $t_{1/f}$ for the initial concentration to decrease by a fraction $1/f$ is given by

$$t_{1/f} = \frac{\ln f}{k_r}$$

The rate constant can therefore be calculated directly from a measurement of this fractional life or the half-life of the reaction.

Example 2.1

The following results were obtained for the decomposition of glucose in aqueous solution.

Glucose concentration/mmol dm^{-3}	56.0	55.3	54.2	52.5	49.0
Time/min	0	45	120	240	480

Show that the reaction is first order and calculate the rate constant for the process and the half-life for glucose under these conditions.

From the above data, $a = 56.0$ mmol dm^{-3} and the glucose concentration readings correspond to $a - x$ readings in equation 2.8, provided the reaction is first order.

log$_{10}$ [(a − x)/mmol dm^{-3}]	1.748	1.743	1.734	1.719	1.690
t/min	0	45	120	240	480

A plot of log$_{10}$ (a − x) against t is given in figure 2.2.

Since the graph is a straight line, the reaction is first order, and

$$\text{slope} = -\frac{k_r}{2.303} = -1.18 \times 10^{-4} \text{ min}^{-1}$$

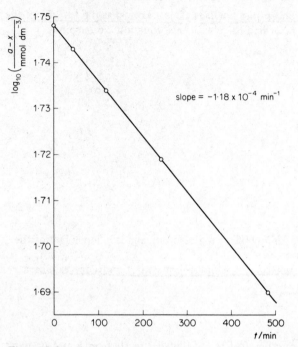

Figure 2.2   *First-order plot for the decomposition of glucose in aqueous solution*

that is

$$k_r = 2.72 \times 10^{-4} \text{ min}^{-1}$$

From equation 2.9

$$t_{0.5} = \frac{0.693}{k_r} = \frac{0.693}{1.18 \times 10^{-4}} \text{ min}$$
$$= 5.87 \times 10^3 \text{ min}$$

2.4   Second-order integrated rate equations

2.4.1   *Reaction involving two reactants*
Consider the reaction

$$A + B \rightarrow \text{products}$$

Let the initial concentrations of A and B be a and b respectively. Let x be the decrease in the concentration of A and B in time t. At time t the concentration of A and B is $a - x$ and $b - x$, respectively. The rate equation

$$-\frac{d[A]}{dt} = -\frac{d[B]}{dt} = k_r[A][B]$$

becomes

$$\frac{dx}{dt} = k_r(a - x)(b - x)$$

or

$$\frac{dx}{(a - x)(b - x)} = k_r \, dt$$

Expressing as partial fractions gives

$$\frac{1}{a - b}\left[\frac{1}{b - x} - \frac{1}{a - x}\right] dx = k_r \, dt$$

On integrating

$$k_r t = \frac{\ln(a - x) - \ln(b - x)}{a - b} + \text{constant}$$

When $t = 0$, $x = 0$, and

$$\text{constant} = \frac{\ln a/b}{a - b}$$

giving

$$k_r t = \frac{1}{a - b} \ln\left[\frac{b(a - x)}{a(b - x)}\right]$$

or

$$k_r = \frac{2.303}{t(a - b)} \log_{10}\left[\frac{b(a - x)}{a(b - x)}\right] \tag{2.10}$$

2.4.2 Reaction involving a single reactant or reaction between two reactants with equal initial concentrations

For the reaction

$$2A \rightarrow \text{products}$$

or the reaction

$$A + B \rightarrow \text{products}$$

where the initial concentrations of A and B are equal, let the initial concentration be a. Equation 2.2 becomes

$$\frac{dx}{dt} = k_r(a - x)^2$$

or

$$\frac{dx}{(a - x)^2} = k_r \, dt$$

On integrating

$$k_r t = \frac{1}{a - x} + \text{constant}$$

Since $x = 0$ at $t = 0$, constant $= -1/a$ and

$$k_r t = \frac{1}{a - x} - \frac{1}{a}$$

or

$$k_r = \frac{1}{at}\left(\frac{x}{a - x}\right) \tag{2.11}$$

2.4.3   Determination of second-order rate constants

(i) Substitution method
The rate constant is calculated by substitution of the experimental values of $a - x$ and $b - x$ obtained at different times t into equation 2.10. If the calculated values of k_r are constant within experimental error, the reaction is assumed to be second order and the average value of k_r gives the second-order rate constant.

(ii) Graphical method
For a second-order reaction of type 2.4.1, equation 2.10 can be rearranged to give

$$\log_{10}\left(\frac{a - x}{b - x}\right) = -\log_{10}\frac{b}{a} + \frac{k_r(a - b)}{2.303}t \tag{2.12}$$

A plot of $\log_{10}(a - x)/(b - x)$ against t will be linear with a slope equal to $k_r(a - b)/2.303$ from which k_r is determined.

Example 2.2

The following kinetic data were obtained by Slater (*J. chem. Soc.,* **85** (1904), 286) for the reaction between sodium thiosulphate and methyl iodide at $25°C$, concentrations being expressed in arbitrary units.

Time/min	0	4.75	10	20	35	55	∞
$[Na_2S_2O_3]$	35.35	30.5	27.0	23.2	20.3	18.6	17.1
$[CH_3I]$	18.25	13.4	9.9	6.1	3.2	1.5	0

Show that the reaction is second order.

If the reaction is second order, equation 2.12 is obeyed, $a - x$ and $b - x$ being the concentrations of $Na_2S_2O_3$ and CH_3I, respectively, at time t.

$\log_{10}(a - x)/(b - x)$	0.287	0.357	0.436	0.580	0.802	1.093
t/min	0	4.75	10	20	35	55

A plot of $\log_{10}(a - x)/(b - x)$ against t is given in figure 2.3. Since the plot is linear, the reaction is second order.

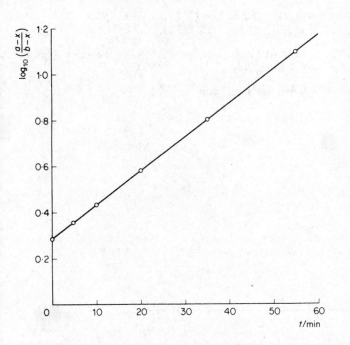

Figure 2.3 *Second-order plot for the reaction between sodium thiosulphate and methyl iodide*

For a second-order reaction of type 2.4.2 where a is equal to b or the reaction involves only one reactant of initial concentration a, it can be seen that a plot of $1/(a - x)$ against t will be linear as in figure 2.4 and that the second-order rate constant is equal to the slope.

Example 2.3

The saponification of ethyl acetate in sodium hydroxide solution at $30°\,C$

$$CH_3CO_2C_2H_5 + NaOH \rightarrow CH_3CO_2Na + C_2H_5OH$$

was studied by Smith and Lorenson (*J. Am. chem. Soc.*, **61** (1939), 117). The initial concentration of ester and alkali were both 0.05 mol dm^{-3}, and the decrease in ester concentration x was measured at the following times.

$10^3\,x/\text{mol dm}^{-3}$	5.91	11.42	16.30	22.07	27.17	31.47	36.44
Time/min	4	9	15	24	37	53	83

Calculate the rate constant for the reaction.

If the reaction is second order, equation 2.11 will be obeyed.

$\text{dm}^3 \text{ mol}^{-1}/(a - x)$	22.7	25.9	29.7	35.8	43.8	53.9	73.8
t/min	4	9	15	24	37	53	83

A plot of $1/(a - x)$ against t is given in figure 2.4. Since the graph is linear, the reaction is second order and

$$\text{slope} = k_r = 0.640 \text{ dm}^3 \text{ mol}^{-1} \text{ min}^{-1}$$

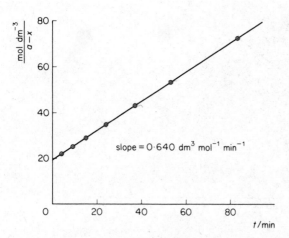

Figure 2.4   *Second-order plot for the reaction between ethyl acetate and sodium hydroxide at $30°C$*

(iii) Fractional-life method

The fractional-life method is only applicable to second-order reactions of type 2.4.2. Since the half-life, for example, is the time required for the initial concentration to decrease from a to $a/2$, equation 2.11 with $x = a/2$ becomes

$$t_{0.5} = \frac{1}{k_r a} \frac{(a/2)}{(a/2)} = \frac{1}{k_r a} \tag{2.13}$$

Therefore, for this type of second-order reaction, the half-life is inversely proportional to the initial concentration, and the rate constant is determined directly from a measurement of the half-life.

If the half-life is measured in two experiments in which two different initial concentrations, a_1 and a_2, are used, then the relationship

$$(t_{0.5})_1/(t_{0.5})_2 = a_2/a_1$$

is obeyed for second-order reactions.

The fractional-life method is applicable to a reaction of any order provided all the reactants have the same initial concentration. In general the half-life of a reaction of order n is related to the initial concentration by

$$t_{0.5} \propto \frac{1}{a^{n-1}}$$

or

$$t_{0.5} = \frac{\text{constant}}{a^{n-1}}$$

Taking logarithms

$$\log_{10} t_{0.5} = (1 - n) \log_{10} a + \log_{10} \text{constant}$$

A plot of $\log_{10} t_{0.5}$ against $\log_{10} a$ is linear with slope equal to $1 - n$. It is possible to derive the rate constant from the intercept.

Alternatively, if $(t_{0.5})_1$ is the half-life for an initial concentration a_1 and $(t_{0.5})_2$ is the half-life when the initial concentration is a_2, then

$$(t_{0.5})_1/(t_{0.5})_2 = (a_2/a_1)^{n-1} \tag{2.14}$$

and taking logarithms

$$\log_{10} (t_{0.5})_1/(t_{0.5})_2 = (n - 1) \log_{10} a_2/a_1$$

from which n can be obtained.

Example 2.4

When the concentration of A in the simple reaction $A \rightarrow B$ was changed from 0.51 mol dm^{-3} to 1.03 mol dm^{-3}, the half-life dropped from 150 seconds to 75 seconds at 25°C. What is the order of the reaction and the value of the rate constant?

Substitution in equation 2.14 gives

$$\frac{150}{75} = \left(\frac{1.03}{0.5}\right)^{n-1}$$

or

$$\log_{10} 2 \approx (n-1) \log_{10} 2$$

that is

$$n = 2$$

Since the reaction is second order, the rate constant is given by equation 2.13; that is

$$t_{0.5} = \frac{1}{k_r a}$$

Therefore

$$k_r = \frac{1}{0.51 \times 150} \text{ dm}^3 \text{ mol}^{-1} \text{ s}^{-1}$$
$$= 1.31 \times 10^{-2} \text{ dm}^3 \text{ mol}^{-1} \text{ s}^{-1}$$

Example 2.5
The reaction

$$SO_2Cl_2 \rightarrow SO_2 + Cl_2$$

is a first-order gas reaction with a rate constant of $2.0 \times 10^{-5} \text{ s}^{-1}$ at 320°C. What percentage of SO_2Cl_2 is decomposed on heating at 320°C for 90 min?

For a first-order reaction, equation 2.7 is obeyed; that is

$$k_r t = 2.303 \log_{10}\left(\frac{a}{a-x}\right)$$

This equation becomes

$$k_r t = 2.303 \log_{10}\left(\frac{1}{1-y}\right)$$

where y is the fraction of SO_2Cl_2 decomposed in time t. Substituting the appropriate numerical values

$$2.0 \times 10^{-5} \times 90 \times 60 = 2.303 \log_{10}\left(\frac{1}{1-y}\right)$$

so that

$$\frac{1}{1-y} = 1.114$$

that is

$$y = 0.102$$

Therefore, SO_2Cl_2 is 10.2 per cent decomposed.

(iv) Isolation method

This method is used to determine the order with respect to each reactant by controlling the reaction conditions such that only one of the reactants changes with time for any one set of experiments. The method can be illustrated by reference to the oxidation of iodides by hydrogen peroxide in acid solution

$$H_2O_2 + 2I^- + 2H_3O^+ \rightarrow I_2 + 4H_2O$$

The rate is given by

$$v = \frac{d[I_2]}{dt} = k_r [H_2O_2]^a [I^-]^b [H_3O^+]^c$$

where a, b and c is the order with respect to each of the reactants and k_r is the rate constant. In the presence of a large excess of acid, $[H_3O^+]$ is effectively constant, and if thiosulphate is added to reconvert the iodine formed back to iodide, $[I^-]$ is also constant. Under these conditions

$$v = k_1 [H_2O_2]^a$$

and the reaction is first order if $a = 1$ or second order if $a = 2$, etc. The rate constant k_1 can be evaluated by one of the methods described previously. The experiment is repeated under conditions where H_2O_2 is in excess and b determined. By a similar approach c can also be calculated.

2.5   Third-order integrated rate equations

Consider the general third-order reaction

$$A + B + C \rightarrow products$$

If a, b and c are the initial concentrations of A, B and C respectively, and x is the decrease in their concentration in time t, the rate equation is

$$\frac{dx}{dt} = k_r(a-x)(b-x)(c-x) \tag{2.15}$$

For the simplest case, where the initial concentrations are all equal to a, equation 2.15 becomes

$$\frac{dx}{dt} = k_r(a - x)^3$$

or

$$\frac{dx}{(a - x)^3} = k_r\, dt$$

On integrating

$$\frac{1}{2(a - x)^2} = k_r t + \text{constant}$$

Since $x = 0$ at $t = 0$, constant $= 1/2a^2$, and

$$k_r t = \frac{1}{2(a - x)^2} - \frac{1}{2a^2} \tag{2.16}$$

The rate constant can be determined by substitution of the appropriate experimental data in this equation or from a plot of $1/(a - x)^2$ versus t, or from a measurement of the half-life, since substitution of $x = a/2$ when $t = t_{0.5}$ in equation 2.16 gives

$$k_r t_{0.5} = \frac{3}{2a^2}$$

2.6   Opposing reactions

Consider two opposing first-order reactions

$$A \underset{k_{-1}}{\overset{k_1}{\rightleftharpoons}} B$$

where k_1 and k_{-1} denote the rate constants of the forward and reverse reactions.

Let a be the initial concentration of A and x be the decrease in concentration of A in time t, and x_e be the decrease in concentration of A at equilibrium.

The concentrations of A and B are therefore

	Initially	At time t	At equilibrium
A	a	$a - x$	$a - x_e$
B	0	x	x_e

The rate of reaction is given by

$$-\frac{d[A]}{dt} = \frac{dx}{dt} = k_1(a-x) - k_{-1}x$$

At equilibrium, this becomes

$$0 = k_1(a - x_e) - k_{-1}x_e$$

that is

$$k_{-1} = \frac{k_1(a - x_e)}{x_e}$$

or

$$k_1a = x_e(k_1 + k_{-1}) \tag{2.17}$$

Therefore at time t

$$\frac{dx}{dt} = k_1(a-x) - \frac{k_1(a - x_e)x}{x_e}$$

$$= \frac{k_1a(x_e - x)}{x_e}$$

Integrating

$$\int_0^x \frac{dx}{x_e - x} = \frac{k_1a}{x_e} \int_0^t dt$$

that is

$$\ln\left(\frac{x_e}{x_e - x}\right) = \frac{k_1at}{x_e}$$

or

$$k_1 = \frac{2.303x_e}{at} \log_{10}\left(\frac{x_e}{x_e - x}\right)$$

Substituting in equation 2.17 gives

$$k_1 + k_{-1} = \frac{2.303}{t} \log\left(\frac{x_e}{x_e - x}\right)$$

which is comparable to the expression for a simple first-order process.

TABLE 2.1 SUMMARY OF ELEMENTARY RATE LAWS

Order	Rate law in differential form	Rate law in integrated form	Typical units of k	Half-life proportional to
0	$\dfrac{dx}{dt} = k$	$kt = x$	$\text{mol dm}^{-3}\,\text{s}^{-1}$	a^1
1	$\dfrac{dx}{dt} = k(a-x)$	$kt = \ln \dfrac{a}{(a-x)}$	s^{-1}	$a^0(=1)$
2	$\dfrac{dx}{dt} = k(a-x)^2$	$kt = \dfrac{x}{a(a-x)}$	$\text{dm}^3\,\text{mol}^{-1}\,\text{s}^{-1}$	a^{-1}
2	$\dfrac{dx}{dt} = k(a-x)(b-x)$	$kt = \dfrac{1}{a-b}\ln\dfrac{b(a-x)}{a(b-x)}$	$\text{dm}^3\,\text{mol}^{-1}\,\text{s}^{-1}$	–
3	$\dfrac{dx}{dt} = k(a-x)^3$	$kt = \dfrac{1}{2(a-x)^2} - \dfrac{1}{2a^2}$	$\text{dm}^6\,\text{mol}^{-2}\,\text{s}^{-1}$	a^{-2}

Problems

1. Calculate the rate constant for the gas-phase reaction between hydrogen and iodine at 681 K if the rate of loss of iodine was $0.192\ \text{N m}^{-2}\,\text{s}^{-1}$ when the initial pressure of I_2 was $823\ \text{N m}^{-2}$ and the initial pressure of H_2 was $10\,500\ \text{N m}^{-2}$. If the iodine pressure is unchanged and the initial hydrogen pressure is $39\,500\ \text{N m}^{-2}$, what is the rate of the reaction?

2. The following kinetic data were obtained for the reaction between nitric oxide and hydrogen at $700°C$

$$2NO + H_2 \rightarrow N_2 + 2H_2O$$

Initial conc./mol dm^{-3}		Initial rate/mol dm^{-3} s^{-1}
NO	H$_2$	
0.025	0.01	2.4×10^{-6}
0.025	0.005	1.2×10^{-6}
0.0125	0.01	0.6×10^{-6}

Deduce:
(a) the order of the reaction with respect to each reactant;
(b) the rate constant of the reaction at $700°C$.

3. The catalysed decomposition of H_2O_2 in aqueous solution was followed by titrating samples with $KMnO_4$ at various time intervals to determine undecomposed H_2O_2

Time/min	5	10	20	30	50
Volume of $KMnO_4/cm^3$	37.1	29.8	19.6	12.3	5.0

Show graphically that the reaction is first order and determine the rate constant.

4. The following data were obtained for the hydrolysis of a sugar in aqueous solution at $23°C$

Time/min	0	60	130	180
Sugar concentration/mol dm^{-3}	1.000	0.807	0.630	0.531

Show that the reaction is first order and calculate the rate constant for the hydrolysis.

5. During the hydrolysis of an alkyl bromide

$$RBr + OH^- \rightarrow ROH + Br^-$$

in aqueous alcohol, it was found that it took 47×10^3 s to liberate 0.005 mol dm^{-3} of free bromide ion when the initial reactant concentrations were both 0.01 mol dm^{-3}. The same percentage conversion took 4.7×10^3 s with initially 0.1 mol dm^{-3} of both reactants. Calculate the rate constant of the hydrolysis reaction.

[University of Manchester BSc (1st year) 1972]

6. The reaction between triethylamine and methyl iodide gives a quaternary amine

$$(C_2H_5)_3N + CH_3I \rightarrow CH_3(C_2H_5)_3NI$$

At $20°C$ with initial concentrations $[amine]_0 = [CH_3I]_0 = 0.224$ mol dm^{-3} in carbon tetrachloride solution, the reaction was followed by determining unreacted amine potentiometrically giving the following results

Time/min	10	40	90	150	300
Amine concentration/mol dm^{-3}	0.212	0.183	0.149	0.122	0.084

Show that the rate constant is second order overall and calculate the rate constant.

[University of Liverpool BSc Honours (Part 1) 1972]

7. The neutralisation reaction of nitroethane in aqueous solution proceeds according to the rate equation

$$-\frac{d[OH]}{dt} = -\frac{d[C_2H_5NO_2]}{dt} = k[C_2H_5NO_2][OH^-]$$

Experiments at $0°C$ with initial concentrations of both reactants equal to 0.01 mol dm^{-3} give a value of 150 s for the reaction half-life. Calculate the corresponding rate constant at $0°C$.

8. Two substances A and B undergo a bimolecular reaction step. The following table gives the concentrations of A at various times for an experiment carried out at a constant temperature of $17°C$

$10^4 [A]/\text{mol dm}^{-3}$	10.00	7.94	6.31	5.01	3.98
Time/min	0	10	20	30	40

The initial concentration of B is 2.5 mol dm^{-3}. Calculate the second-order rate constant for the reaction.

[University of Manchester, B.Sc. (1st year) 1967]

9. In the second-order reaction between isobutyl bromide and sodium ethoxide in ethanol at $95°C$, the initial concentration of bromide was $0.0505 \text{ mol dm}^{-3}$ and of ethoxide was $0.0762 \text{ mol dm}^{-3}$. The decrease x in the concentration of both the reactants was measured as follows

$10^3 x/\text{mol dm}^3$	0	5.9	10.7	16.6	23.0	27.7	33.5
Time/min	0	5	10	17	30	40	60

Calculate the rate constant for the reaction.

10. The following results were obtained for the decomposition of ammonia on a heated tungsten surface

Initial pressure/torr	65	105	150	185
Half-life/s	290	460	670	820

Determine the order of the reaction.

3 EXPERIMENTAL METHODS FOR THE DETERMINATION OF REACTION RATES

The rate of a chemical reaction is defined as the rate of decrease of the concentration of reactants or the rate of increase of the concentration of products. To measure the rate, it is necessary to follow the change in the concentration of a reactant or product with time by a convenient method. Numerous methods are available and the choice of a suitable technique depends on the answer to some of the following questions:

1. Is the half-life of the reaction sufficiently long to allow a conventional method to be used?
2. Is the reaction carried out in the gas phase or in solution?
3. Can the reactants or products be readily analysed?
4. Is there some physical property of the system, for example electrical conductivity, optical rotation, absorbance, viscosity, which changes during the reaction?
5. Is at least one of the products a gas?

The rates of chemical reactions are very sensitive to changes in temperature. It is normally necessary to carry out kinetic experiments in a vessel kept in a thermostatically controlled bath the temperature of which does not vary by more than $0.1°C$. A laboratory stop clock is sufficiently accurate to measure reaction times, although it is often difficult to start the clock at the exact zero time. For reactions in solution, the two reactants are mixed together as quickly as possible. The mixing must be rapid relative to the half-life of the reaction, and problems can arise when a reactant is added from a pipette. If a high-temperature decomposition or isomerisation reaction is studied, the reactant must first be kept at a temperature at which it is stable, and then heated as rapidly as possible to the reaction temperature. Since some reaction occurs before this temperature is reached, it is difficult to record an accurate zero time reading in such reactions.

Glass reaction vessels are usually quite satisfactory, since this surface is inert to most laboratory chemicals. In gas-phase investigations, reactions often take place on the vessel walls. These surface reactions are reduced by washing the vessel walls or coating them with an inert substance, but reproducibility is difficult to attain. A reaction is shown to be homogeneous if its rate is unaffected by a change in the surface-to-volume ratio. If the reaction rate changes when some glass chippings are added to the vessel, the reaction is assumed to be heterogeneous.

The usual aim of a kinetic experiment is to determine the form of the rate equation and a value for the rate constant. This can be achieved by two methods corresponding to the two forms of the rate equation given in chapter 2.

(i) Differential method
This method is a direct method, since values of dc/dt are determined directly from plots of concentration versus time, the tangents to the curve at any time t giving the rate at that time.

(ii) Integration method
In this method the experimental variation of c with t is compared to one of the integrated rate equations derived in chapter 2. If the rate data plotted graphically satisfy the first-, second- or third-order equation, the order is known, and the rate constant is determined as described previously. This method is not very satisfactory for reactions with non-integral orders, since data plotted for a reaction of order 1.2 do not differ much from a first-order graphical plot.

3.1 Differential methods

3.1.1 *The initial rate method*
Consider a reaction

$$A \rightarrow B$$

The rate v is given by

$$v = k_r c_A{}^n$$

where k_r is the rate constant, c_A is the initial concentration of A and n is the order. Taking logarithms

$$\log_{10} v = \log_{10} k_r + n \log_{10} c_A$$

Graphs of concentration against time are plotted for a number of initial concentrations c_1, c_2, c_3, etc., and the tangent at the start of the reaction is drawn as in figure 3.1. This corresponds to the initial rate for that particular initial concentration. When the rate is measured at the start of a reaction, it can be assumed that complications caused by the presence of secondary reactions do not occur. The logarithm of the initial rate is plotted against the logarithm of the initial concentration as in figure 3.2, and the rate constant and order determined from the intercept and slope respectively.

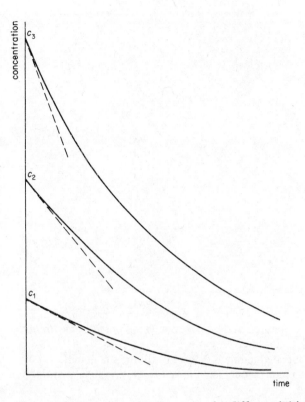

Figure 3.1 *Concentration–time curve for different initial concentrations*

One advantage of this method over integration methods is that it is not dependent on a knowledge of the order of the reaction. The main disadvantage is the difficulty of drawing tangents accurately. This practical problem can be overcome by a measurement of the time for a small fraction of reactant to change to a product.

In the following reaction in solution

$$S_2O_8^{2-} + 2I^- \rightarrow 2SO_4^{2-} + I_2 \tag{1}$$

the initial rate is determined by introducing a small known concentration of sodium thiosulphate with starch into the system so that the reaction

$$2S_2O_3^{2-} + I_2 \rightarrow S_4O_6^{2-} + 2I^- \tag{2}$$

occurs between the added thiosulphate ions and the iodine liberated in the reaction (1) under study. When all the thiosulphate ions are consumed in

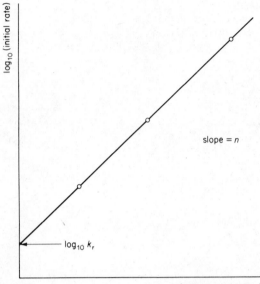

Figure 3.2 *Plot of $\log_{10}$ (initial rate) against $\log_{10}$ (initial concentration)*

reaction (2), the liberated iodine is detected with the starch indicator by the appearance of a blue colour. The time Δt to the appearance of this blue colour is measured.

The rate of reaction is given by

$$-\frac{d[S_2O_8^{2-}]}{dt} = k_r[S_2O_8^{2-}][I^-]$$

and for small changes in $[S_2O_8^{2-}]$ this equation can be expressed in the form

$$-\frac{\Delta[S_2O_8^{2-}]}{\Delta t} = k_r[S_2O_8^{2-}]_0[I^-]_0 \tag{3.1}$$

where $-\Delta[S_2O_8^{2-}]$ represents the concentration of $[S_2O_8^{2-}]$ used up during the reaction in time Δt, and $[S_2O_8^{2-}]_0$ and $[I^-]_0$ are the initial reactant concentrations of the persulphate and iodide ions, respectively. An accurate value of the rate constant can be determined from these measurements provided $\Delta[S_2O_8^{2-}]$ is kept to less than 1 per cent of $[S_2O_8^{2-}]_0$, that is the rate is measured for 1 per cent of reaction. The order of the above reaction can be confirmed from a plot of $1/\Delta t$ against $[S_2O_8^{2-}]_0$ with $[I^-]_0$ constant, and a plot of $1/\Delta t$ against $[I^-]$ with $[S_2O_8^{2-}]_0$ constant, since both these plots are linear.

3.2 Integration methods

Integration methods fall basically into two types: (i) methods involving the withdrawal of a sample for analysis and (ii) methods in which the reaction mixture is analysed or a physical property measured *in situ*, that is continuously.

3.2.1 Sampling methods

This method normally involves the removal of a sample of the reaction mixture, and the sample is cooled or diluted to stop further reaction within the sample. The latter is then analysed for one or more of the reactants or products by a suitable method. The procedure is repeated at different reaction times.

Alternatively, if a reaction mixture contains a gas or volatile solvent, the reaction can be carried out in a sealed tube. This provides a convenient method to study reactions at high temperatures. The reactants are sealed in the tube at room temperature and quickly brought up to the reaction temperature. The sealed tube is removed at a certain time and cooled in order to stop the reaction. It is broken and the contents are analysed. Since some reaction occurs before the reaction temperature is reached, it is advisable to use two tubes, one being a control tube which is removed and cooled as soon as the reaction temperature is reached. The concentration of reactants in this tube corresponds to the initial concentrations, and this constitutes the zero-time reading.

The following techniques are used to analyse the samples.

(i) Titration

The original investigation of Bodenstein of the reaction between hydrogen and iodine

$$H_2 + I_2 \rightarrow 2HI$$

was carried out in a sealed tube. A mixture of hydrogen and iodine vapour was heated in a quartz tube at 700 K, and after a certain time broken under aqueous alkali at room temperature. The excess hydrogen was collected and measured, and the solution analysed volumetrically for iodine and iodide concentrations. The experiment was repeated at different reaction times in other tubes.

The acid-catalysed hydrolysis of an ester is often studied by this technique

$$CH_3CO_2CH_3 + H_2O \rightarrow CH_3CO_2H + CH_3OH$$

When the reaction is carried out in excess dilute acid, it is first order, the rate depending only on the ester concentration

$$\text{rate} = k_r [CH_3CO_2CH_3]$$

where k_r is a pseudo-first-order rate constant. Methyl acetate and excess dilute hydrochloric acid are mixed in a flask fitted with a soda lime tube to exclude CO_2 from the atmosphere. Small samples are withdrawn from the mixture at frequent time intervals and immediately diluted to stop further reaction. Titration of these samples against standard alkali using phenolphthalein as an indicator enables the liberated acetic acid (which does not change the H^+ ion concentration) to be measured.

The initial concentration of ester is proportional to $T_\infty - T_0$, where T_∞ is the titre at a time corresponding to complete reaction, and T_0 is the titre at zero time, that is the titre for the dilute HCl alone. The change in ester concentration in time t (corresponding to $a - x$ in the integrated rate equation) is proportional to $T_\infty - T_t$ where T_t is the titre at any time t. The first-order rate constant is obtained from the relationship

$$k_r = \frac{2.303}{t} \log_{10} \left(\frac{T_\infty - T_0}{T_\infty - T_t} \right)$$

or

$$\log_{10}(T_\infty - T_t) = \log_{10}(T_\infty - T_0) - \frac{k_r}{2.303} t \tag{3.2}$$

A plot of $\log_{10}(T_\infty - T_t)$ against t is linear with slope equal to $-k_r/2.303$.

Example 3.1

A certain amount of methyl acetate was hydrolysed in the presence of an excess of 0.05 mol dm^{-3} hydrochloric acid at $25°C$. When 25 cm^3 aliquots of reaction mixture were removed and titrated with NaOH solution, the volume V of alkali required for neutralisation after time t were as follows.

t/min	0	21	75	119	∞
V/cm^3	24.4	25.8	29.3	31.7	47.2

Show that the reaction is first order and calculate the rate constant.

If the reaction is first order, equation 3.2 is obeyed.

$(T_\infty - T_t)$/cm^3	22.8	21.4	17.9	15.5
$\log_{10}[(T_\infty - T_t)/cm^3]$	1.358	1.330	1.253	1.190

A plot of $\log_{10}(T_\infty - T_t)$ against t is given in figure 3.3.

Since the graph is linear, the reaction is first order, and

$$\text{slope} = -\frac{k_r}{2.303} = -1.46 \times 10^{-3} \text{ min}^{-1}$$

that is

$$k_r = 3.36 \times 10^{-3} \text{ min}^{-1}$$

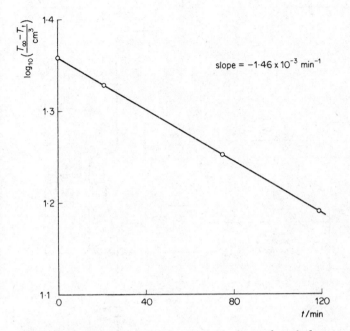

Figure 3.3   *First-order plot for the hydrolysis of methyl acetate in excess dilute HCl at 25°C*

(ii) Gas chromatography

In the last twenty years, the development of gas chromatographic techniques has facilitated the study of the kinetics of many gas-phase reactions. This analytical technique is extremely sensitive and can be used to analyse a large number of chemical compounds simultaneously. With gas–solid chromatography or gas–liquid chromatography it is possible to analyse the products of a very complex reaction and identify the minor as well as the major reaction products. The quantitative estimation of the yield of each product often enables the relative rates of many of the individual steps in the reaction to be determined. A typical gas–liquid chromatographic apparatus set up for sampling products of a gas-phase reaction is shown in figure 3.4.

(iii) Spectroscopy and mass spectrometry

These techniques are particularly valuable in flow systems where low concentrations of reactants or products can be measured *in situ*. However, even in conventional kinetics, the analysis of reaction mixtures by infrared or ultraviolet spectroscopy and mass spectroscopy, particularly if the latter is linked to a gas chromatograph, has proved to be a useful method for sample analysis.

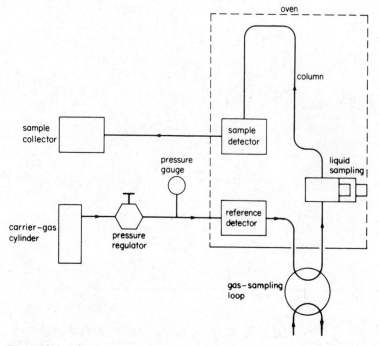

Figure 3.4 *Typical gas–liquid chromatograph*

3.2.2 Continuous methods

A number of kinetic methods are available in which a physical property of the reaction system is measured at time intervals throughout the reaction. In this way the reaction mixture is not disturbed by the removal of samples and the reaction is allowed to proceed to completion. The following techniques are most commonly used as continuous methods.

(i) Electrical conductivity method

The electrical conductivity method is useful for the study of reactions in which ions, particularly H^+ ions or OH^- ions, which have relatively high ionic conductivities, are involved. In dilute solution, the replacement of one ion by another of different ionic conductivity will be proportional to the rate of change in the concentration of the reacting ion.

For example, in the alkaline hydrolysis of an ester

$$CH_3CO_2C_2H_5 + OH^- \rightarrow C_2H_5OH + CH_3CO_2^-$$

the conductivity of the solution decreases as the highly conducting hydroxyl ions are replaced by acetate ions. This reaction is second order and for equal initial concentrations a of reactants, the rate equation is given by equation (2.11)

$$k_r = \frac{1}{at}\left(\frac{x}{a-x}\right)$$

Let κ_0 be the initial conductivity of the solution, κ_t the conductivity after time t and κ_∞ the conductivity when the reaction has gone to completion. Since the decrease in concentration x in time t is proportional to $\kappa_0 - \kappa_t$ and the concentration $a - x$ at time t is proportional to $\kappa_t - \kappa_0$, the equation becomes

$$k_r = \frac{1}{at}\left(\frac{\kappa_0 - \kappa_t}{\kappa_t - \kappa_\infty}\right)$$

Rearrangement gives

$$\kappa_t = \frac{1}{k_r a}\left(\frac{\kappa_0 - \kappa_t}{t}\right) + \kappa_\infty \tag{3.3}$$

A plot of κ_t versus $(\kappa_0 - \kappa_t)/t$ is linear with slope equal to $1/k_r a$.

(ii) Optical rotation method
This method is confined to substances that are optically active. The muta-rotation of glucose, catalysed by acids or bases, is a first-order reaction. The α-glucose has a specific rotation of $+110°$, while β-glucose has a specific rotation of $19°$. On standing, either isomer yields an equilibrium mixture of α and β forms with $[\alpha]_0 = 52.5°$. Therefore the angle through which plane-polarised light rotates as the isomerisation

$$\alpha\text{-glucose} \underset{k_{-1}}{\overset{k_1}{\rightleftharpoons}} \beta\text{-glucose}$$

occurs, provides a measure of the rate of reaction. The angle of rotation is measured in a polarimeter at time intervals of about a minute for half an hour.

It has been shown (page 23) that the rate constant for this reaction is given by

$$k_r t = 2.303 \log_{10}\left(\frac{x_e}{x_e - x}\right)$$

where $k_r = k_1 + k_{-1}$ and x and x_e are the concentrations of β isomer at time t and at equilibrium, respectively. If α_0 is the initial angle of rotation, α_∞ is the final angle of rotation and α_t is the angle of rotation at time t, x_e is proportional to $\alpha_0 - \alpha_\infty$ and $x_e - x$ is proportional to $\alpha_t - \alpha_\infty$, giving

$$k_r t = 2.303 \log_{10}\left(\frac{\alpha_0 - \alpha_\infty}{\alpha_t - \alpha_\infty}\right)$$

or

$$\log_{10}(\alpha_t - \alpha_\infty) = \log_{10}(\alpha_0 - \alpha_\infty) - \frac{k_r t}{2.303}$$

Therefore a plot of $\log_{10}(\alpha_t - \alpha_\infty)$ against t is linear with slope equal to $-k_r/2.303$.

(iii) Spectrophotometric method

If a solution obeys Beer's Law, the absorbance of a reactant or product is proportional to its concentration. Therefore, provided a region of the spectrum is chosen where the absorption is due to a single compound in the reaction mixture, the reaction can be followed spectrophotometrically by a measurement of the absorbance at that wavelength as a function of time.

The reaction of cinnamal chloride with ethanol yields 1-chloro-3-ethoxy-3-phenyl-1-propene

$$C_6H_5CH\!=\!CHCHCl_2 + C_2H_5OH \rightarrow C_6H_5CHCH\!=\!CHCl + HCl$$

(I)
$$\underset{OC_2H_5}{|}$$
(II)

Reactant I absorbs strongly at 260 nm, while product II does not absorb in that region, so that the rate is measured from the change in absorbance A with time.

Let the absorbance be A_0 at the start of the reaction and A_t at time t. Therefore, A_0 is proportional to the initial concentration of reactants a and A_t is proportional to the concentration of reactants $a - x$ at time t. Provided excess alcohol is present, the reaction is first order and obeys the equation

$$k_r t = 2.303 \log_{10}\left(\frac{a}{a - x}\right) = 2.303 \log_{10}\left(\frac{A_0}{A_t}\right)$$

A plot of $\log A_t$ against t is linear with slope equal to $-k_r/2.303$.

When the reaction mixture contains two absorbing species, it is often difficult to select a convenient wavelength for the measurements. The cis-trans isomerisation of bisethylenediaminedichlorocobalt(III) chloride in methanol is a first-order reaction but both the isomers absorb in the visible region. However, the absorption spectrum of the two compounds (figure 3.5) shows that at 540 nm the cis isomer has an absorption maximum while

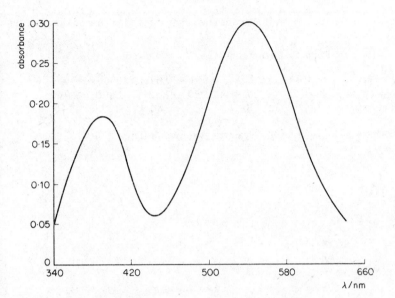

Figure 3.5   *Absorption spectrum of the chromium III–EDTA complex at pH 4.1*

the *trans* isomer has an absorption minimum. The absorbance of the reaction mixture A_t is measured as a function of time, and the infinity time reading A_∞ corresponds to complete conversion to the *trans* isomer. The difference $A_t - A_\infty$ is therefore proportional to the concentration of *cis* isomer at any time t, and the first-order rate equation is given by

$$k_r t = 2.303 \log_{10}\left(\frac{A_0 - A_\infty}{A_t - A_\infty}\right)$$

where A_0 is the absorbance at zero time. Therefore

$$\log_{10}(A_t - A_\infty) = \log_{10}(A_0 - A_\infty) - \frac{k_r t}{2.303} \tag{3.4}$$

and a plot of $\log_{10}(A_t - A_\infty)$ versus t is linear with slope equal to $-k_r/2.303$.

Example 3.2
The rate of disappearance of the absorption peak at 540 nm in the above reaction as a function of time was found to be as follows

Absorbance	0.119	0.108	0.096	0.081	0.071	0.060	0.005
Time/min	0	20	47	80	107	140	∞

Show that the reaction is first order and calculate the half-life of the reaction.

If the reaction is first order, equation 3.4 is obeyed

$A_t - A_\infty$	0.114	0.103	0.091	0.076	0.066	0.055
$\log_{10}(A_t - A_\infty)$	$\bar{1}.057$	$\bar{1}.013$	$\bar{2}.959$	$\bar{2}.881$	$\bar{2}.820$	$\bar{2}.740$
Time/min	0	20	47	80	107	140

A plot of $\log_{10}(A_t - A_\infty)$ against t is shown in figure 3.6.

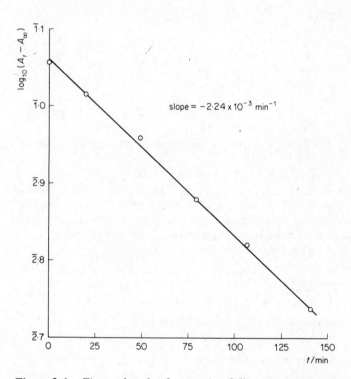

Figure 3.6   *First-order plot for reaction followed spectrophotometrically*

The graph is linear and therefore the reaction is first order. From the graph

$$\text{slope} = \frac{-k_r}{2.303} = -\,2.24 \times 10^{-3} \text{ min}^{-1}$$

that is

$$k_r = 5.16 \times 10^{-3} \text{ min}^{-1}$$

$$t_{0.5} = \frac{0.693}{5.16 \times 10^{-3}} = 134 \text{ min}$$

(iv) Dilatometric method

If a chemical reaction involves a change in volume, the volume change is directly proportional to the extent of reaction. In the acid-catalysed hydrolysis of acetal in excess water

$$CH_3CH(OC_2H_5)_2 + H_2O \xrightarrow{H^+} CH_3CHO + 2C_2H_5OH$$

an increase in volume occurs and this can be measured by a dilatometer. This consists of a reaction vessel to which is connected a fine, uniform capillary with a calibrated scale. The change in the level of the liquid in the capillary during the reaction is read off using a cathetometer, and the volume change determined.

If V_0, V_∞ and V_t correspond to the initial volume, final volume and volume at time t, respectively, the first-order rate equation is given by

$$\log_{10}(V_\infty - V_t) = \log_{10}(V_\infty - V_0) - \frac{k_r t}{2.303}$$

A plot of $\log_{10}(V_\infty - V_t)$ versus t is linear with slope equal to $-k_r/2.303$.

(v) Gas evolution method

Consider a reaction in solution in which one of the products is a gas. Benzenediazonium chloride decomposes in aqueous solution at room temperature liberating nitrogen according to the equation

$$C_6H_5\overset{+}{N}_2Cl^- + H_2O \rightarrow C_6H_5OH + HCl + N_2$$

In excess water the reaction is first order and obeys the first-order rate equation

$$k_r t = 2.303 \log_{10}\left(\frac{a}{a-x}\right) \tag{3.5}$$

Let V_∞ be the volume of nitrogen evolved when the reaction is allowed to go to completion, and let V_t be the volume evolved in time t. Since a is proportional to V_∞ and x is proportional to V_t, equation 3.5 becomes

$$k_r t = 2.303 \log_{10}\left(\frac{V_\infty}{V_\infty - V_t}\right)$$

or

$$\log_{10}(V_\infty - V_t) = \log V_\infty - \frac{k_r t}{2.303} \qquad (3.6)$$

A plot of $\log_{10}(V_\infty - V_t)$ against t is linear with slope equal to $-k_r/2.303$.

The apparatus is shown in figure 3.7. The solution is placed in the reaction vessel and allowed to come to thermal equilibrium in a thermostat. The gas evolved is usually allowed to replace water from a gas burette, and the volume of gas evolved is measured at convenient time intervals. The V_∞ reading is best obtained by immersing the reaction vessel in a beaker of hot water until no more gas evolves. The vessel is cooled to the temperature of the experiment, and the volume reading recorded.

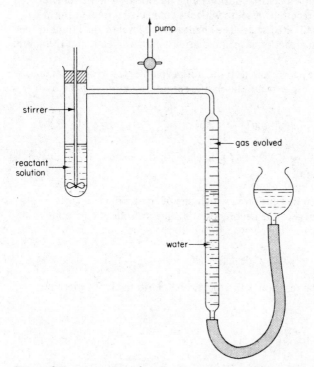

Figure 3.7   *Apparatus for the determination of reaction rate by the gas evolution method*

Example 3.3

The following data refer to the decomposition of benzenediazonium chloride in solution at 50°C

N_2 evolved/cm^3	19.3	26.0	32.6	36.0	41.3	45.0	48.4	58.3
Time/min	6	9	12	14	18	22	26	∞

Find the order of the reaction and the rate constant.

If the reaction is first order, equation 3.6 will be obeyed.

$V_\infty - V_t$/cm^3	39.0	32.3	25.7	22.3	17.0	13.3	9.9
$\log_{10}[(V_\infty - V_t)/cm^3]$	1.592	1.509	1.410	1.348	1.230	1.124	0.996
t/min	6	9	12	14	18	22	26

A plot of $\log_{10}(V_\infty - V_t)$ against t is given in figure 3.8.

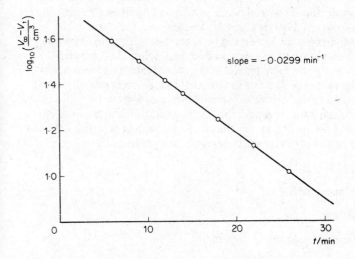

Figure 3.8 *First-order plot for the decomposition of benzenediazonium chloride at $50°C$*

Since the plot is linear, the reaction is first order, and

$$\text{slope} = -\frac{k_r}{2.303} = -0.0299 \text{ min}^{-1}$$

that is

$$k_r = 0.0651 \text{ min}^{-1}$$

3.3 Gas-phase reactions

Reactions in which all the reactants and products are gases can be conveniently studied by a measurement of the change in total pressure in a system kept at constant volume. This method is therefore restricted to reactions in which there is an increase or decrease in the number of molecules during the reaction. The pressure change is measured on a manometer (provided the volume change in the manometer itself is negligible compared to the total volume) or by a spiral gauge. Alternatively pressure gauges are available, which produce an electrical signal, which is led to a recorder to give a direct pressure–time plot.

The thermal decomposition of ethane at 856 K is a first-order reaction, which gives ethylene and hydrogen as products

$$C_2H_6 \rightarrow C_2H_4 + H_2$$

The rate of this reaction could be determined by taking out small samples of the reaction mixture to analyse for ethane and hydrogen by gas–solid chromatography, as described previously. However, the reaction can be studied by following the increase in total pressure with time, since one molecule decomposes to give two product molecules. In all gas-phase reactions followed by this technique, it is necessary to derive a relationship between the partial pressure of the reactant (which decreases with time) and the total pressure (which in this case increases with time).

Let p_0 be the initial pressure; this corresponds to the pressure of ethane at $t = 0$. Let y be the decrease in the ethane pressure at any time t. Then at time t, the partial pressures of the reactant and products are

Ethane	$p_{C_2H_6} = p_0 - y$
Ethylene	$p_{C_2H_4} = y$
Hydrogen	$p_{H_2} = y$

The total pressure p at time t is given by

$$p = p_{C_2H_6} + p_{C_2H_4} + p_{H_2}$$
$$= p_0 - y + y + y$$
$$= p_0 + y$$

Therefore

$$y = p - p_0$$

and the partial pressure of ethane $p_{C_2H_6}$ at time t is given by

$$p_{C_2H_6} = p_0 - (p - p_0)$$
$$= 2p_0 - p$$

For the first-order rate equation

$$k_r t = 2.303 \log_{10}\left(\frac{a}{a - x}\right)$$

the initial concentration a is proportional to p_0 and the concentration $a - x$ is proportional to $2p_0 - p$. Therefore, the rate constant k_r is determined from the equation

$$k_r t = 2.303 \log_{10}\left(\frac{p_0}{2p_0 - p} \right)$$

or

$$\log_{10}(2p_0 - p) = \log_{10} p_0 - \frac{k_r t}{2.303} \tag{3.7}$$

A plot of $\log_{10}(2p_0 - p)$ against t is linear with slope equal to $-k_r/2.303$.

The form of the relationship depends on the overall stoichiometry of the reaction, and, in each case, it is necessary to derive the appropriate expression that relates the decrease in the pressure of reactant with the change in the total pressure.

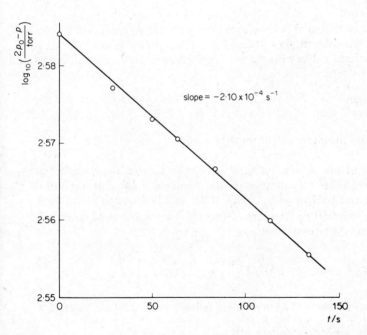

Figure 3.9   *First-order plot for the thermal decomposition of ethane at 856 K*

Example 3.4

The following changes in total pressure p at constant volume with time t was observed for the above reaction at 856 K

p/torr	384	390	394	396	400	405	408
t/s	0	29	50	64	84	114	134

Determine the rate constant at this temperature.

Since the above reaction is first order, equation 3.7 is obeyed.

$2p_0 - p$/torr	384	378	374	372	368	363	360
$\log_{10}[(2p_0 - p)/\text{torr}]$	2.584	2.577	2.573	2.571	2.566	2.560	2.556
t/s	0	29	50	64	84	114	134

A plot of $\log_{10}(2p_0 - p)$ against t is plotted in figure 3.9. From the graph

$$\text{slope} = -\frac{k_r}{2.303} = -2.10 \times 10^{-4} \text{ s}^{-1}$$

$$k_r = 4.83 \times 10^{-4} \text{ s}^{-1}$$

Problems

1. The oxidation of potassium iodide by potassium persulphate in excess iodide was followed by withdrawing samples at various times and titrating the liberated iodine with sodium thiosulphate solution. The following results were obtained at 25°C

Time/min	0	5	10	15	20	25	30	∞
Titre/cm^3	0.00	5.80	9.80	12.60	14.50	15.80	16.80	18.80

Calculate the rate constant for the reaction.

2. The hydrolysis of ethyl nitrobenzoate by aqueous sodium hydroxide was followed at 25°C by titration of the hydroxide at different stages of the reaction with 0.01 mol dm^{-3} HCl. If the initial concentrations of both reactants is 0.01 mol dm^{-3} show that the reaction is second order and calculate the rate constant.

Time/s	95	140	222	334	805
Titre/cm^3	9.3	9.0	8.5	7.9	6.1

3. The hydrolysis of t-butyl iodide follows first-order kinetics at 26°C

$$\text{Bu}^t\text{I} + \text{H}_2\text{O} \rightarrow \text{Bu}^t\text{OH} + \text{HI}$$

The conductivity of the solution was measured with time as follows

Time/min			0	2	4	6	8	10	17	22	28	∞
10^6 x conductivity/												
Ω^{-1} m^{-1}			5.5	13.0	20.0	26.0	31.0	36.0	46.0	51.5	56.0	65.0

Calculate the rate constant of the reaction.

4. The following results were obtained for the reversible *cis-trans* isomerisation of stilbene at 280°C

Time/s	0	1830	3816	7260	12 006	∞
% *cis* isomer	100	88.1	76.3	62.0	48.5	17

Calculate the rate constants of the forward and reverse reactions.

5. The optical rotations α for the mutarotation of α-glucose at 20°C are as follows

α/degrees	20.26	18.92	16.82	15.22	14.06	13.18	10.60
Time/min	10	20	40	60	80	100	equilibrium

Show that the reaction is first order and calculate the rate constant $k_1 + k_{-1}$.

6. The rate of a reaction was followed by measuring the absorbance of a solution at various times

Time/min	0	18	57	130	240	337	398
Absorbance	2.19	2.06	1.83	1.506	1.198	1.051	0.980

Assuming the Beer–Lambert law is obeyed, show that the reaction is first order and determine the rate constant.

7. Reaction of diacetone alcohol with alkali to give acetone produces a volume change, which is measured on a cathetometer as the reaction proceeds. From the following data obtained by G. Akerlof (*J. Am. chem. Soc.*, **49** (1927), 2955) with 5 per cent diacetone alcohol by volume in a KOH solution of concentration 2 mol dm^{-3}, calculate the order and the rate constant.

Time/s	0	24.4	35.0	48.0	64.8	75.8	89.4	106.6	133.4	183.6	∞
Cathetometer reading	8.0	20.0	24.0	28.0	32.0	34.0	36.0	38.0	40.0	42.0	43.3

8. A solution of hydrogen peroxide (15 cm^3) is catalytically decomposed by colloidal silver into oxygen and water; complete decomposition of this solution gives 6.18 cm^3 of oxygen at s.t.p., and the volume of oxygen collected over a period of time t is tabulated.

Time/min	2	4	6	8	14
Volume O_2/cm^3	1.24	2.36	3.36	3.98	5.23

Determine the order of this reaction and calculate the rate constant.

9. The decomposition of ethylene oxide

$$C_2H_4O \rightarrow CH_4 + CO$$

at 687 K gave the following results

Total pressure/torr	116.5	122.6	125.7	128.7	133.2	141.4
Time/min	0	5	7	9	12	18

Show that the reaction is first order and calculate the rate constant.

10. The gas-phase decomposition of di-t-butyl peroxide is a first-order reaction given by

$$(CH_3)_3COOC(CH_3)_3 \rightarrow 2CH_3COCH_3 + C_2H_6$$

The following results were obtained for the total pressure p measured at constant volume at $147°C$ at times t

t/min	0	6	10	14	22	30	38	46
p/torr	179.5	198.6	210.5	221.2	242.3	262.1	280.1	297.1

Calculate the rate constant for the reaction.

Further reading

Reviews

L. Batt. *Comprehensive Chemical Kinetics*, **1** (1969), 1.

Books

A. M. James. *Practical Physical Chemistry*, Churchill, London (1967).
H. Melville and B. G. Gowenlock. *Experimental Methods in Gas Reactions*, Macmillan, London (1964).
J. H. Purnell. *Gas Chromatography*, Wiley, New York (1962).
J. M. Wilson, R. J. Newcombe, A. R. Denaro and R. M. W. Rickett. *Experiments in Physical Chemistry* (2nd Edition), Pergamon, Oxford (1968).

4 DEPENDENCE OF RATE ON TEMPERATURE

4.1 Arrhenius equation

The rate equation of the form

$$\text{rate} = k_r[A]^{n_1}[B]^{n_2} \ldots \tag{4.1}$$

expresses the dependence of reaction rate on the concentrations of the reactants. However, the rate of a reaction varies greatly with temperature, since for a typical process the rate doubles or trebles for a rise in temperature of 10°C.

In equation 4.1 the concentration terms and the order are not sensitive to changes in temperature, and it is the rate constant k_r which is the temperature-dependent term.

It was found empirically that the rate constant k_r varies with temperature according to the relationship

$$\log_{10} k_r = b - \frac{a}{T}$$

where a and b are constants and T is the absolute or thermodynamic temperature. It was shown by van't Hoff and Arrhenius that the theoretical basis for this law is the relationship between the equilibrium constant K_c and temperature known as the van't Hoff isochore

$$\frac{d \ln K_c}{dT} = \frac{\Delta E}{RT^2} \tag{4.2}$$

where K_c is the equilibrium constant in terms of concentration and ΔE is the energy change.

Consider a reaction

$$A + B \underset{k_{-1}}{\overset{k_1}{\rightleftharpoons}} C + D$$

The rate of the forward reaction is $k_1[A][B]$, and the rate of the reverse reaction is $k_{-1}[C][D]$, where k_1 and k_{-1} are the rate constants for the forward and reverse reactions, respectively.

At equilibrium

$$k_1[A][B] = k_{-1}[C][D]$$

and the equilibrium constant is given by

$$K_c = \frac{[C][D]}{[A][B]} = \frac{k_1}{k_{-1}}$$

Equation 4.2 becomes

$$\frac{d \ln k_1}{dT} - \frac{d \ln k_{-1}}{dT} = \frac{\Delta E}{RT^2}$$

and can be expressed as two equations

$$\frac{d \ln k_1}{dT} = \frac{E_1^{\ddagger}}{RT^2} + I$$

and

$$\frac{d \ln k_{-1}}{dT} = \frac{E_{-1}^{\ddagger}}{RT^2} + I$$

where $\Delta E = E_1^{\ddagger} - E_{-1}^{\ddagger}$ and I is an integration constant. Arrhenius found that for a number of reactions, I was equal to zero and formulated his law as

$$\boxed{\frac{d \ln k_r}{dT} = \frac{E^{\ddagger}}{RT^2}} \qquad (4.3)$$

where k_r is the rate constant and $E^{\ddagger}$ is the energy term, which for reasons given in the next section is known as the *activation energy*. This relationship is called the *Arrhenius equation* and is one of the most important equations in kinetics.

An alternative form of the Arrhenius equation is

$$k_r = A \exp(-E^{\ddagger}/RT) \qquad (4.4)$$

where A is a constant known as the *frequency factor*.

4.2   Activation energy

Since I was found equal to zero, it follows that the rate of the forward reaction depends on $E_1^{\ddagger}$ and the rate of the reverse reaction depends on $E_{-1}^{\ddagger}$. This implies that the reaction path from A + B to C + D involves an energy change $E_1^{\ddagger}$ while the reverse reaction involves an energy change $E_{-1}^{\ddagger}$, the difference in energy being ΔE.

These conditions are satisfied if the reaction proceeds via an intermediate state, which has energy $E_1^\ddagger$ greater than the initial state and energy $E_{-1}^\ddagger$ greater than the final state. This is illustrated in the energy diagram shown in figure 4.1.

The intermediate state is known as the activated state. This state is in equilibrium with A and B and is said to be an *activated complex*. The super-script ‡ is used to indicate the activated state. Molecules A and B must acquire energy $E_1^\ddagger$ before they can form a complex and hence C + D; this

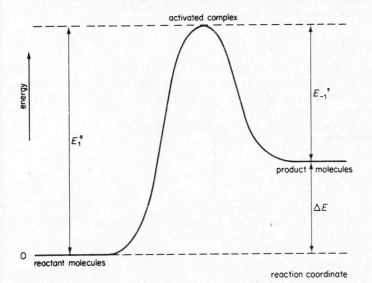

Figure 4.1   *Potential energy–reaction co-ordinate diagram for an endo-thermic reaction*

energy is termed the *activation energy* and is written as $E_1^\ddagger$. This is there-fore the minimum energy that A and B must acquire before reaction takes place to give C + D.

It can be seen that for the reverse reaction C and D must acquire energy $E_{-1}^\ddagger$ before forming the activated complex and hence A + B. The activation energy of the reverse step is therefore $E_{-1}^\ddagger$.

Modern theory of chemical change postulates that the additional energy required for thermal reaction is attained by collisions between the mole-cules. It will be seen that only a small number of molecules may possess this energy of activation and it is only these molecules that can react in the chemical sense to give products.

Alternatively the energy of activation can be thought of as a potential energy barrier. Only molecules with sufficient energy to reach the top of the barrier and form the activated complex react. It is easy to visualise that the lower this barrier (that is, the smaller the activation energy) the greater the number of activated molecules, and the faster is the rate of reaction.

4.2.1 Determination of activation energy

It can be seen that the rate constant for any reaction depends on two factors: (a) the frequency of collisions between the reactant molecules; (b) the value of the activation energy. The Arrhenius equation expressed in the form given in equation 4.4

$$k_r = A \exp(-E^{\ddagger}/RT) \tag{4.4}$$

where $E^{\ddagger}$ is the activation energy and A is the frequency factor, shows this dependence on (a) and (b). The frequency factor has the same dimensions as the rate constant and is related to the frequency of collisions between the reactant molecules.

The logarithmic form of equation 4.4 is given by

$$\ln k_r = \ln A - E^{\ddagger}/RT \tag{4.5}$$

or

$$\log_{10} k_r = \log_{10} A - \frac{E^{\ddagger}}{2.303RT} \tag{4.6}$$

Equation 4.5 can also be obtained by integration of equation 4.3 provided the integration constant is put equal to $\ln A$.

It can be seen that the activation energy of a reaction can be determined if the rate constant is measured at a number of different temperatures. A plot of $\log_{10} k_r$ against $1/T$ is linear if equation 4.6 is obeyed, and the slope is equal to $-E^{\ddagger}/2.303R$.

Example 4.1

Some values of the rate constant for the alkaline hydrolysis of ethyl iodide over the temperature range $20°C$–$80°C$ are as follows

$10^3 k_r/dm^3$ mol^{-1} s^{-1}	0.100	0.335	1.41	3.06	8.13	21.1	50.1
Temperature/°C	20	30	40	50	60	70	80

Calculate the activation energy of the reaction.

Figure 4.2 shows that the rate constant increases exponentially as the temperature increases. It can be seen that when the temperature is increased from 313 K to 323 K, the rate constant approximately doubles.

$\log_{10}(k_r/dm^3$ mol^{-1} s^{-1})	$\bar{4}.000$	$\bar{4}.525$	$\bar{3}.015$	$\bar{3}.485$	$\bar{3}.910$	$\bar{2}.325$	$\bar{2}.700$
Temperature/K	293	303	313	323	333	343	353
10^3 K$/T$	3.413	3.300	3.195	3.096	3.003	2.915	2.833

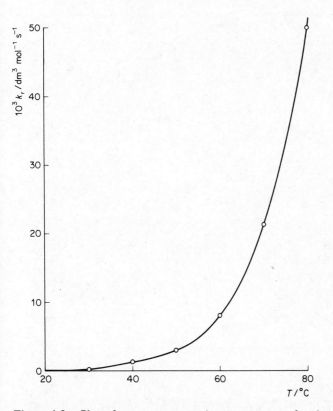

Figure 4.2 *Plot of rate constant against temperature for the alkaline hydrolysis of ethyl iodide*

A plot of $\log_{10} k_r$ against $1/T$ is shown in figure 4.3. From the slope of the graph

$$-E^{\ddagger}/2.303R = -4.70 \times 10^3 \text{ K}$$

that is

$$E^{\ddagger} = 2.303 \times 8.314 \times 4.70 \times 10^3 \text{ J mol}^{-1}$$
$$= 90.0 \text{ kJ mol}^{-1}$$

Alternatively, if the rate constants k_1 and k_2 are measured at two temperatures T_1 and T_2, respectively, the activation energy can be calculated as follows. From equation 4.3

$$\frac{\text{d} \ln k_r}{\text{d}T} = \frac{E^{\ddagger}}{RT^2}$$

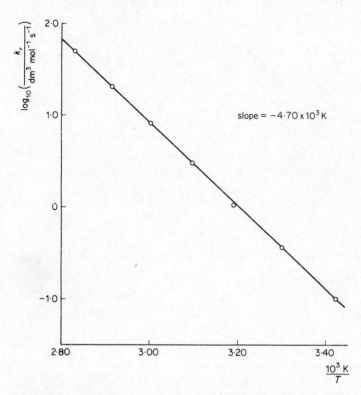

Figure 4.3 *A plot of $\log_{10} k_r$ against $1/T$ for the alkaline hydrolysis of ethyl iodide*

Integrating

$$\int_{k_1}^{k_2} d \ln k_r = \frac{E^\ddagger}{R} \int_{T_1}^{T_2} \frac{dT}{T^2}$$

gives

$$\left[\ln k_r\right]_{k_1}^{k_2} = - \frac{E^\ddagger}{R} \left[\frac{1}{T} \right]_{T_1}^{T_2}$$

or

$$\ln \frac{k_2}{k_1} = - \frac{E^\ddagger}{R} \left[\frac{1}{T_2} - \frac{1}{T_1} \right]$$

that is

$$\log_{10} \frac{k_2}{k_1} = \frac{E^{\ddagger}}{2.303R} \left[\frac{1}{T_1} - \frac{1}{T_2} \right]$$

or

$$\log_{10} \frac{k_2}{k_1} = \frac{E^{\ddagger}}{2.303R} \left[\frac{T_2 - T_1}{T_1 T_2} \right] \tag{4.7}$$

The frequency factor can be determined from the intercept of a $\log_{10} k_r$ against $1/T$ plot, but this normally requires a very long extrapolation of the graph to $1/T = 0$. Therefore, a more accurate value can be calculated from a known value of k_r at a temperature T by substitution in equation 4.6.

Example 4.2
The initial stage of the reaction between gaseous ammonia and nitrogen dioxide follows second-order kinetics. Given that the rate constant at 600 K is 0.385 dm^3 mol^{-1} s^{-1} and at 716 K is 16.0 dm^3 mol^{-1} s^{-1}, calculate the activation energy and the frequency factor.

Substitution in equation 4.7 gives

$$\log_{10} \left(\frac{16.0}{0.385} \right) = \frac{E^{\ddagger}}{2.303 \times 8.314} \left(\frac{716 - 600}{716 \times 600} \right)$$

giving

$$E^{\ddagger} = \frac{1.619 \times 2.303 \times 8.314 \times 716 \times 600}{116} \text{ J mol}^{-1}$$
$$= 114.8 \text{ kJ mol}^{-1}$$

Using the value of k_2 at 716 K in equation 4.6 gives

$$\log_{10} (16.0/dm^3 \text{ mol}^{-1} s^{-1}) = \log_{10} A - \frac{114\,800}{2.303 \times 8.314 \times 716}$$

giving

$$A = 3.8 \times 10^9 \text{ dm}^3 \text{ mol}^{-1} s^{-1}$$

4.3 The activated complex
The activated complex is thought to be the intermediate state of all chemical reactions. In order to predict theoretically the rate of a reaction, it is necessary to postulate the configuration of the activated complex or the transition state.

Consider a reaction between an atom A and a diatomic molecule BC

$$A + BC \rightarrow AB + C$$

For reaction to occur, atom A must approach the molecule BC. As A gets closer to BC, electronic interaction occurs and the potential energy increases. This increase continues until the configuration A—B—C is formed such that either AB + C or BC + A is obtained. This intermediate configuration is the activated complex for this process represented by

$$A + BC \rightarrow A—B—C \rightarrow AB + C$$

When atom C separates to give molecule AB (or if the reaction merely reverses, to the original A + BC) the potential energy decreases again. The activated state is therefore the position of maximum potential energy along the reaction path. A potential energy diagram for this process is shown in figure 4.4.

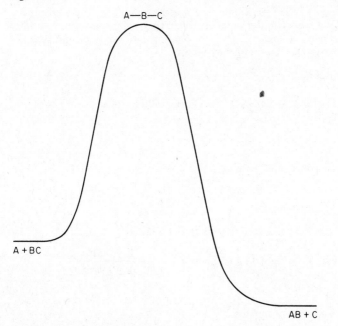

Figure 4.4   *Potential-energy barrier for the reaction $A + BC \rightarrow AB + C$*

The change in the potential energy when the activated state is formed from the initial state is $E_1^{\ddagger}$, the activation energy for the forward process. Similarly the activation energy for the reverse process

$$AB + C \rightarrow A + BC$$

is $E_{-1}^{\ddagger}$. The difference is equal to the total energy change ΔE, that is

$$\Delta E = E_1^{\ddagger} - E_{-1}^{\ddagger}$$

or in this case, where there is no change in the number of molecules on either side of the chemical equation, at constant pressure, the enthalpy change ΔH

$$\Delta H = E_1^{\ddagger} - E_{-1}^{\ddagger}$$

If $E_1^{\ddagger} < E_{-1}^{\ddagger}$, ΔH is negative as in figure 4.4 and the reaction is exothermic. If $E_1^{\ddagger} > E_{-1}^{\ddagger}$, ΔH is positive and the reaction is endothermic. For endothermic reactions, the activation energy cannot be less than the ΔH of the reaction and the enthalpy change is therefore the minimum possible value for the activation energy. Such processes tend to be slow except at high temperatures.

Many theoretical calculations have been carried out on the reaction between hydrogen atoms and *para* hydrogen, which forms an equilibrium mixture of *ortho* and *para* hydrogen

$$H + H_2(para) \rightarrow H_2(ortho) + H$$

This is a thermoneutral reaction, that is, $\Delta H = 0$ and $E_1^{\ddagger} = E_{-1}^{\ddagger}$. Quantum mechanical calculations have shown that the activated complex in this system is symmetrical, the $H^{\alpha}-H^{\beta}$ bond distance being equal to the $H^{\beta}-H^{\gamma}$ bond distance

$$H^{\alpha} + H^{\beta}-H^{\gamma} \rightarrow H^{\alpha}-H^{\beta}-H^{\gamma} \rightarrow H^{\alpha}-H^{\beta} + H^{\gamma}$$

The reaction between deuterium atoms and molecular hydrogen

$$D + H_2 \rightarrow D-H-H \rightarrow D-H + H$$

and the reaction between a number of halogen atoms X and molecular hydrogen

$$X + H_2 \rightarrow X-H-H \rightarrow X-H + H$$

have also been extensively studied. For these few reactions, Eyring and Polanyi were able to calculate the course of a reaction in terms of the potential energy as a function of the two bond distances in BC and AB as they change during the reaction.

A typical reaction pathway for the reaction

$$A + BC \rightarrow A-B-C \rightarrow AB + C$$

is illustrated in figure 4.5. The potential energy curves for the diatomic molecules BC and AB have minima at points P and Q respectively. One possible reaction pathway is PRQ. The molecule BC decomposes into atoms B and C which react with A in a three-atom process to give the products

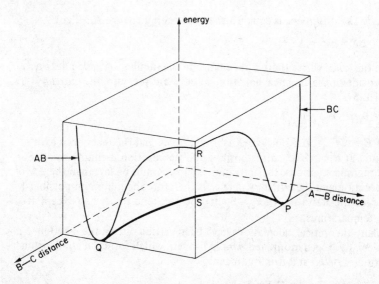

Figure 4.5   *Potential-energy diagram for a linear A—B—C molecule*

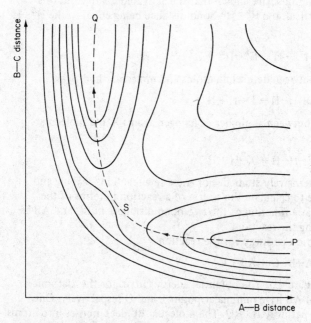

Figure 4.6   *Potential-energy contour diagram for linear A—B—C molecule*

AB and C. This pathway can be represented by

$$A + BC \rightarrow A + B + C \rightarrow AB + C$$

The energy required for this pathway is at least equal to the bond dissociation energy of BC. If BC is *ortho*-H_2, the activation energy would be about 435 kJ mol^{-1}, and the process would be very slow.

The calculations showed that the reaction pathway is represented by PSQ in figure 4.5.

This requires much less energy and is consistent with the experimental activation energy of 35 kJ mol^{-1}. The reaction pathway can be represented by a contour diagram of figure 4.5, which is shown in figure 4.6. The contour lines indicate the energy levels. The reaction pathway proceeds up a valley from P to the saddle point or col S, and then down the valley to Q. This is the most economical pathway in terms of energy. The col, being the point of maximum energy on the reaction pathway, represents the transition state. The height of the col above the energy of P represents the activation energy.

Problems

1. The following first-order rate constants k_r were obtained for the thermal decomposition of ethane

$10^5 k_r/s^{-1}$	2.5	4.7	8.2	12.3	23.1	35.3	57.6	92.4	141.5
Temperature/K	823	833	843	853	863	873	883	893	903

From these values determine the activation energy and the pre-exponential factor for this decomposition. State clearly the units in which these quantities were measured.

[Brunel University, B.Tech. (Year 2) 1972]

2. The rate constants k_r for the reaction

$$2HI \rightarrow H_2 + I_2$$

at different temperatures are

$k_r/dm^3 mol^{-1} s^{-1}$	T/K
3.11×10^{-7}	556
1.18×10^{-6}	575
3.33×10^{-5}	629
8.96×10^{-5}	647
1.92×10^{-4}	666
5.53×10^{-4}	683
1.21×10^{-3}	700

Calculate the activation energy and the frequency factor of the reaction.

3. The rate constant for the decomposition of nitrogen dioxide is $5.22 \times 10^{-5} \, dm^3 \, mol^{-1} s^{-1}$ at 592 K and $17.00 \times 10^{-5} \, dm^3 \, mol^{-1} s^{-1}$ at 627 K. Calculate the activation energy for the reaction.

4. For a certain first-order reaction, the time required to reduce the initial reactant concentration to one-half is 5000 s at 325 K and 1000 s at 335 K. Calculate the rate constant at both temperatures and hence determine the activation energy.

[Brunel University, B.Tech. (Part 1) 1972]

5. In a certain reaction, the rate constant at $35°C$ is double its value at $25°C$. Calculate the activation energy.

6. The decomposition of compound A in solution is a first-order process with an activation energy of $52.3 \, kJ \, mol^{-1}$. A 10 per cent solution of A is 10 per cent decomposed in 10 min at $10°C$. How much decomposition would be observed with a 20 per cent solution after 20 min at $20°C$?

7. The Arrhenius equations for the rate of decomposition of dibutyl mercury and diethyl mercury are

$$k_r/s^{-1} = 10^{15.2} \exp\left(-\frac{193 \, kJ \, mol^{-1}}{RT}\right)$$

and

$$k_r/s^{-1} = 10^{14.1} \exp\left(-\frac{180 \, kJ \, mol^{-1}}{RT}\right)$$

respectively. Find the temperature at which the rate constants are equal.

5 THEORY OF REACTION RATES

5.1 Collision theory

The kinetics of both gas-phase and liquid-phase reactions have been studied extensively for many years. Although experimentally more difficult, the study of gas-phase reactions has led to a satisfactory theory of reaction rates for homogeneous gas-phase reactions.

The first theory of reaction rates postulated that for two gases A and B to react, it was necessary for a collision between A and B molecules to occur. On collision some bonds were broken and others formed, resulting in new molecules being produced. The rate of reaction therefore depends on the rate of collisions between the reactant molecules.

Consider a reaction between two molecules of a single gas A to give products C and D

$$2A \rightarrow C + D$$

From the kinetic theory of gases the collision rate Z_{AA}, that is the number of collisions per unit volume per unit time, is given by

$$Z_{AA} = 2n^2\sigma^2 \left(\frac{\pi RT}{M} \right)^{1/2} \tag{5.1}$$

where n is the concentration of A expressed as molecules per unit volume of gas, σ is the collision diameter, R is the gas constant, T is the thermodynamic temperature and M is the molar mass.

For a reaction between 2 unlike gases A and B to give C and D by

$$A + B \rightarrow C + D$$

the corresponding collision rate Z_{AB} is given by

$$Z_{AB} = n_A n_B \sigma_{AB}^2 \left(\frac{8\pi RT}{\mu} \right)^{1/2} \tag{5.2}$$

where n_A and n_B are the concentrations of A and B, respectively, expressed in molecules per unit volume, σ_{AB} is the mean collision diameter given by

$$\sigma_{AB} = \frac{\sigma_A + \sigma_B}{2}$$

and μ is the reduced mass, that is

$$\mu = \frac{m_A m_B}{m_A + m_B}$$

Calculation of the collision rate for gases at ambient temperatures and pressures shows that very few chemical reactions occur at every collision. For HI molecules at 700 K and atmospheric pressure, $Z = 10^{34}\,\text{m}^{-3}\,\text{s}^{-1}$, but the reaction rate for the decomposition of HI

$$2HI \rightarrow H_2 + I_2 \tag{1}$$

is about $10^{20}\,\text{m}^{-3}\,\text{s}^{-1}$. This means that only one collision in 10^{14} is effective in producing the above reaction.

The basis of simple collision theory is that for reaction to occur, the total relative kinetic energy of molecules A and B on collision must exceed a critical value E_c. In other words, the kinetic energy can be distributed in any way between A and B, but the activated state is only formed when energy greater or equal to a critical energy E_c is attained. It is assumed that the movement of the molecules is confined to a two-dimensional velocity space; that is, this critical energy is distributed over only two degrees of freedom or is expressed by only two squared terms. From the Maxwell–Boltzmann distribution law, the probability P of a molecule of velocity c having simultaneously components of velocity between u and $u + du$ and v and $v + dv$ is given by

$$P = \frac{1}{\pi\alpha^2} \exp\left(-c^2/\alpha^2\right) du\, dv$$

where α is the most probable velocity. If expressed in polar coordinates, the probability of a molecule having velocity between c and $c + dc$ between the limits θ and $\theta + d\theta$ is given by

$$P = \frac{1}{\pi\alpha^2} \exp\left(-c^2/\alpha^2\right)c\, dc\, d\theta$$

Integration between the limits $\theta = 0$ and $\theta = 2\pi$ gives the fraction of the total molecules having velocity between c and $c + dc$, that is

$$\frac{dn_c}{n} = \frac{2}{\alpha^2} \exp\left(-c^2/\alpha^2\right)c\, dc$$

Since $\alpha = 2RT/M$ where M is the molar mass, the fraction of the total molecules having energy between E and $E + dE$ is given by

$$\frac{dn_E}{n} = \frac{1}{RT} \exp\left(-E/RT\right) dE$$

where E is the kinetic energy per mole of all the molecules which have velocity c. The fraction of molecules having energy greater than the critical value of E_c is given by integration

$$\int\limits_{E \geqslant E_c}^{\infty} \frac{dn_E}{n} = \int\limits_{E_c}^{\infty} \frac{\exp(-E/RT)\, dE}{RT}$$
$$= \exp(-E_c/RT)$$

The term $\exp(-E_c/RT)$ is called the Boltzmann factor.

The basis of all modern theories of reaction rates is that only molecules that have energy greater or equal to the activation energy $E^{\ddagger}$ react. The above derivation shows that the fraction of such molecules is $\exp(-E^{\ddagger}/RT)$.

The decomposition of HI by reaction (1) has an activation energy of 184 kJ mol^{-1}. At 700 K, the fraction of activated molecules is therefore given by

$$\exp(-E^{\ddagger}/RT) = \exp\left(-\frac{184\,000}{8.31 \times 700}\right) = 1.85 \times 10^{-14}$$

Thus only one collision in about 10^{14} leads to activation. If the temperature is reduced to 500 K, the corresponding fraction is given by

$$\exp(-E^{\ddagger}/RT) = \exp\left(-\frac{184\,000}{8.31 \times 500}\right) \approx 6 \times 10^{-20}$$

that is, only one collision in about 10^{21} leads to activation. This shows that a rise in temperature from 500 K to 700 K increases the rate by a factor of about 10^7.

It is seen that the higher the value of the activation energy, the smaller the fraction of molecules that are activated and the slower the rate of reaction. Consequently reactions with high activation energies are slow unless carried out at high temperatures.

The rate of reaction v is therefore given by the collision rate multiplied by the fraction of molecules having energy greater than or equal to the activation energy; that is

$$v = 2Z_{AA} \exp(-E^{\ddagger}/RT) \tag{5.3}$$

where the factor of 2 is introduced because two like molecules are involved in each collision.

The rate constant k_r for this reaction is obtained from

$$v = k_r [A]^2 = k_r n^2$$

Therefore

$$k_r = \frac{v}{n^2} = \frac{2Z_{AA}\exp(-E^{\ddagger}/RT)}{n^2}$$

Substituting for Z_{AA} from equation 5.1 gives

$$k_r = 4\sigma^2\left(\frac{\pi RT}{M}\right)^{1/2}\exp(-E^{\ddagger}/RT) \tag{5.4}$$

Example 5.1
Calculate the rate of reaction and the rate constant at 700 K and 1 atm (101.3 kN m^{-2}) pressure for the decomposition of hydrogen iodide assuming its collision diameter is 350 pm.

The number of molecules per unit volume n is given by

$$n = \frac{pN_A}{RT} = \frac{(1.013 \times 10^5) \times 1 \times (6.023 \times 10^{23})}{8.314 \times 700}\ \text{m}^{-3}$$

$$= 1.05 \times 10^{25}\ \text{m}^{-3}$$

Substitution in equation 5.1 gives

$$Z_{AA} = 2n^2\sigma^2\left(\frac{\pi RT}{M}\right)^{1/2}$$

$$= 2 \times (1.05 \times 10^{25})^2 \times (3.5 \times 10^{-10})^2 \times \left(\frac{\pi \times 8.31 \times 700}{127.9 \times 10^{-3}}\right)^{1/2}\ \text{m}^{-3}\text{s}^{-1}$$

$$= 1.02 \times 10^{34}\ \text{m}^{-3}\text{s}^{-1}$$

Therefore from equation 5.3

$$v = 2 \times 1.02 \times 10^{34}\exp\left(\frac{-184\,000}{8.314 \times 700}\right)\text{m}^{-3}\text{s}^{-1}$$

$$= 3.75 \times 10^{20}\ \text{m}^{-3}\text{s}^{-1}$$

$$= 6.23 \times 10^{-4}\ \text{mol m}^{-3}\text{s}^{-1}$$

The rate constant k_r is given by substituting the appropriate numerical values in equation 5.4, giving

$$k_r = 4 \times (3.5 \times 10^{-10})^2\left(\frac{\pi \times 8.31 \times 700}{127.9 \times 10^{-3}}\right)^{1/2}\exp\left(\frac{-184\,000}{8.31 \times 700}\right)$$

$$= 3.40 \times 10^{-30}\ \text{m}^3\text{s}^{-1}$$

Multiplying by the Avogadro constant the molar rate constant is

$$k_r = 3.40 \times 10^{-30} \times 6.02 \times 10^{23}\ \text{m}^3\text{mol}^{-1}\text{s}^{-1}$$

$$= 2.05 \times 10^{-3}\ \text{dm}^3\text{mol}^{-1}\text{s}^{-1}$$

The experimental value of the rate constant obtained for this reaction is $1.57 \times 10^{-3} dm^3 mol^{-1} s^{-1}$, which is in excellent agreement with the calculated value. Therefore there is good agreement between experiment and theory for this reaction.

It can be seen that equation 5.4 is in the same form as the Arrhenius equation given as equation 4.4

$$k_r = A \exp\left(-E^{\ddagger}/RT\right)$$

so that collision theory has derived an expression for the frequency factor A; that is

$$A = 4\sigma^2 \left(\frac{\pi RT}{M}\right)^{1/2}$$

The corresponding collision theory equations for a bimolecular reaction between unlike molecules

$$A + B \rightarrow C + D$$

are

$$v = Z_{AB} \exp\left(-E^{\ddagger}/RT\right) \tag{5.5}$$

Since $v = k_r/n_A n_B$, the rate constant k_r is obtained from this expression and the substitution of equation 5.2 into equation 5.5, giving

$$k_r = \sigma_{AB}^2 \left(\frac{8\pi RT}{\mu}\right)^{1/2} \exp\left(-E^{\ddagger}/RT\right)$$

and the frequency factor is therefore given by

$$A = \sigma_{AB}^2 \left(\frac{8\pi RT}{\mu}\right)^{1/2}$$

5.1.1 Failure of collision theory

It has been shown that for the reaction $2HI \rightarrow H_2 + I_2$ there is good agreement between the experimental rate constant and the value calculated from collision theory. However, this agreement is restricted to a comparatively small number of chemical reactions. In many gas-phase reactions and in most liquid-phase reactions the reaction rate or the rate constant (which is the rate at unit concentration) is very much less than that predicted by simple collision theory. To allow for this discrepancy, equation 5.5 was modified to the form

$$v = PZ_{AB} \exp\left(-E^{\ddagger}/RT\right)$$

or at unit concentration

$$k_r = PZ_{AB} \exp(-E^{\ddagger}/RT) \tag{5.6}$$

where P is the steric factor or probability factor. It can be seen that P is essentially a measure of the discrepancy between the experimental rate constant and the theoretical rate constant. It normally has values from nearly unity in the HI decomposition, to about 10^{-5}, although P factors as low as 10^{-8} have been reported. It is clear that simple collision theory is inadequate for all but a few gas-phase reactions.

All attempts to find a theoretical basis for the steric factor failed and no method was known for predicting its value. It was postulated that for some reactions it is necessary for the reactant molecules to orientate themselves in a certain direction before the transition state was formed. While this is true for some reactions, all attempts to correlate P factors with molecular structure failed. It was also found that for many unimolecular gas-phase reactions and for many reactions in solution, the rate is faster than that predicted; that is, the P factor is greater than unity. Collision theory cannot offer a satisfactory theory to explain this observation.

5.2 Absolute rate theory

Absolute rate theory is also a collision theory that assumes chemical activation occurs by collisions between molecules. It is a development of collision theory that requires a more exact treatment than that provided by kinetic theory. Use is made of statistical thermodynamics to derive an expression for the frequency factor, and this approach, although beyond the scope of this book, has proved to be more precise and productive than kinetic theory.

Absolute rate theory postulates that the rate of a chemical reaction is given by the rate of passage of the activated complex through the transition state. It is assumed that, although the transition state is mechanically unstable, it can be treated thermodynamically in the usual way.

The theory is based on an 'equilibrium hypothesis'. The transition state is made up of complexes that were previously either reactants or products. Equilibrium therefore exists between the reactants A and B and the activated complex $X^{\ddagger}$ and between the products C and D and the activated complex $X^{\ddagger}$

$$A + B \rightleftharpoons X^{\ddagger} \rightleftharpoons C + D$$

On average there is an equal probability that the complex is either moving forwards or backwards. Therefore on average the equilibrium concentration of forward moving complexes is equal to half the total number of complexes at equilibrium. If the products are instantaneously removed from the system, the backward moving complexes are not formed. The hypothesis is that in these circumstances the flow of forward moving complexes is un-affected and that their concentration is the same as in the equilibrium situation. It is, therefore, possible to define $K^{\ddagger}$ as the equilibrium constant for the equilibrium between the reactants and the activated complex $X^{\ddagger}$.

5.3 Thermodynamic formulation of the rate equation

The equilibrium constant K for a chemical reaction is related to the free energy change ΔG by

$$K = \exp(-\Delta G/RT)$$

Since K is given by the ratio of the forward and reverse rate constants k_1/k_{-1}

$$\ln k_1 - \ln k_{-1} = -\Delta G/RT$$

In a similar treatment to that used on page 48 this equation can be expressed as

$$\ln k_1 = -\frac{\Delta G_1}{RT} + \text{constant}$$

and

$$\ln k_{-1} = -\frac{\Delta G_{-1}}{RT} + \text{constant}$$

where ΔG_1 and ΔG_{-1} are the free-energy changes for the forward and reverse reactions, respectively.

For the reaction

$$A + B \rightleftharpoons X^{\ddagger} \rightleftharpoons \text{products}$$

equilibrium exists between the reactants and the activated complex. If $\Delta G^{\ddagger}$ is the difference in free energy between the reactants and the activated state, the rate constant is related in a similar way; that is

$$\ln k_r = -\frac{\Delta G^{\ddagger}}{RT} + \text{constant}$$

Rearranging

$$k_r = \nu \exp(-\Delta G^{\ddagger}/RT)$$

where ν is a constant and $\Delta G^{\ddagger}$ is the free energy of activation.

It can be shown by the use of statistical thermodynamics that the constant ν is equal to kT/h, where k is the Boltzmann constant and h is Planck's constant. Therefore

$$k_r = \frac{kT}{h} \exp\left(-\Delta G^{\ddagger}/RT\right) \qquad (5.7)$$

or

$$k_r = \frac{kT}{h} K^{\ddagger}$$

where $K^{\ddagger}$ represents the equilibrium constant for the equilibrium between the reactant state and the activated state.

Since for a reaction $\Delta G = \Delta H - T\Delta S$, equation 5.7 can be expressed as

$$k_r = \frac{kT}{h} \exp\left(\frac{\Delta S^{\ddagger}}{R}\right) \exp\left(\frac{-\Delta H^{\ddagger}}{RT}\right) \qquad (5.8)$$

where $\Delta H^{\ddagger}$ is the enthalpy of activation and $\Delta S^{\ddagger}$ is the entropy of activation. This equation was derived by Wynne-Jones and Eyring[1].

The activation energy $E^{\ddagger}$ is related to the enthalpy of activation by the equation

$$\Delta H^{\ddagger} = E^{\ddagger} - nRT \qquad (5.9)$$

where n is unity for unimolecular reactions and all liquid-phase reactions, and equal to 2 for bimolecular gas reactions. Substitution of these values for equation 5.9 into equation 5.8 gives the rate equation in terms of the entropy of activation and the energy of activation.

For unimolecular gas reactions and liquid-phase reactions

$$k_r = \frac{kT}{h} \exp\left(1 + \frac{\Delta S^{\ddagger}}{R}\right) \exp\left(-\frac{E^{\ddagger}}{RT}\right) \qquad (5.10)$$

For bimolecular gas reactions

$$k_r = \frac{kT}{h} \exp\left(2 + \frac{\Delta S^{\ddagger}}{R}\right) \exp\left(-\frac{E^{\ddagger}}{RT}\right) \qquad (5.11)$$

Comparison of equation 5.11 with the Arrhenius equation shows that the frequency factor A of a bimolecular reaction is given by

$$A = \frac{kT}{h} \exp\left(2 + \frac{\Delta S^{\ddagger}}{R}\right) \qquad (5.12)$$

and is seen to depend on the entropy of activation.

In general the frequency factor is related to the entropy of activation by the relationship

$$A = \frac{kT}{h} \exp\left(n + \frac{\Delta S^{\ddagger}}{R} \right) \tag{5.13}$$

where n is the molecularity.

Example 5.2
The frequency factor for the decomposition of ethylvinyl ether is $2.7 \times 10^{11}\,\text{s}^{-1}$. Calculate the entropy of activation at $530°\text{C}$.

The units of the frequency factor indicate that the reaction is first order and therefore equation 5.10 will be obeyed. The frequency factor A is given by

$$A = \frac{kT}{h} \exp\left(1 + \frac{\Delta S^{\ddagger}}{R} \right)$$

Taking logarithms

$$\ln A = \ln \frac{kT}{h} + 1 + \frac{\Delta S^{\ddagger}}{R}$$

or

$$\Delta S^{\ddagger} = R \left[2.303 \log_{10} A - 2.303 \log_{10} \left(\frac{kT}{h} \right) - 1 \right]$$

Substituting the appropriate numerical values

$$\Delta S^{\ddagger} = 8.314 \left[2.303 \log_{10} (2.7 \times 10^{11}) \right.$$

$$\left. - 2.303 \log_{10} \left(\frac{1.38 \times 10^{-23} \times 803}{6.626 \times 10^{-34}} \right) - 1 \right] \text{J K}^{-1}\,\text{mol}^{-1}$$

$$= -42.7\ \text{J K}^{-1}\,\text{mol}^{-1}$$

5.4 Entropy of activation

When two reactions of similar activation energy are studied at the same temperature and found to proceed at appreciably different rates, there must be a difference in their entropies of activation. In terms of collision theory, the pre-exponential term PZ is given by

$$PZ = \frac{kT}{h} \exp\left(2 + \frac{\Delta S^{\ddagger}}{R} \right)$$

P factors close to unity indicate that the entropy of activation is approximately zero and the reaction proceeds at a 'normal' rate, or a rate close to that predicted by collision theory. If the P factor is less than unity, the entropy of activation is negative. A rate that is much slower than normal has a low P factor and a high negative $\Delta S^{\ddagger}$. Similarly, if a reaction proceeds at a faster rate than predicted by collision theory or has a P factor greater than unity, absolute rate theory predicts a positive entropy of activation.

The entropy of activation $\Delta S^{\ddagger}$ is equal to the difference in entropy between the reactants and the activated state; that is

$$\Delta S^{\ddagger} = S_{\text{activated state}} - S_{\text{reactants}}$$

Entropy can be understood as a measure of the disorder or randomness of the system. A positive entropy of activation indicates that the activated state is less ordered or has more freedom than the reactants. Conversely, a negative entropy of activation indicates an increase in order or a loss of freedom when the activated state is formed from the reactants.

Consider a bimolecular reaction

$$A + B \rightarrow X^{\ddagger}$$

The activated complex is formed by an association of the reactant molecules A and B. This must of necessity involve a loss of freedom. Two independent molecules lose some translational and rotational freedom when the complex is formed. The greater this loss of freedom, the more negative is the entropy of activation. It would be predicted that the combination of two atoms with no rotational or vibrational degrees of freedom to form a diatomic complex involves only a slight loss of freedom and the entropy of activation is small. This is known as the 'rigid sphere model', since molecules that react with only a small change in freedom or $\Delta S^{\ddagger}$ have rates close to that predicted by collision theory. On the other hand a reaction between two polyatomic molecules with a considerable number of vibrational degrees of freedom to form a single activated complex results in a considerable loss of freedom and a correspondingly high negative entropy of activation. The relationship between the loss of freedom and the reaction rate is illustrated by table 5.1 where the reactions are listed in order of decreasing loss of freedom when the activated complex is formed.

TABLE 5.1

Reaction	$A/\text{dm}^3 \text{ mol}^{-1}\text{s}^{-1}$	$\Delta S_{298}^{\ddagger}/\text{J K}^{-1} \text{ mol}^{-1}$
$H_2 + I_2 = 2HI$	1×10^{11}	8
$2HI = H_2 + I_2$	6×10^{10}	0
$NO + O_3 = NO_2 + O_2$	8×10^8	-33
Dimerisation of 1,3-butadiene	4.7×10^7	-54
Dimerisation of cyclopentadiene	8.5×10^4	-111

It is difficult to visualise a bimolecular reaction in which there is an increase in freedom in forming the activated complex. Consequently, bimolecular reactions with rates greater than normal are rare.

For a unimolecular reaction, the frequency factor is given by

$$A = \frac{kT}{h} \exp\left(1 + \frac{\Delta S^{\ddagger}}{R}\right)$$

A reaction with zero energy of activation has a frequency factor given by $kT/h \exp(1)$, which at 300 K is approximately equal to 10^{13} s^{-1}. Therefore the normal frequency factor for unimolecular reaction is 10^{13} s^{-1}. Unlike bimolecular reactions an inspection of a table of frequency factors for unimolecular reactions shows that many rates are greater than normal; that is, the frequency factors are greater than 10^{13} s^{-1}. Such reactions have a positive entropy of activation. Other unimolecular reactions with frequency factors less than normal have negative entropies of activation.

An isomerisation or decomposition reaction that involves a cyclic reactant molecule usually forms an activated state with a gain of freedom. For example, the isomerisation of cyclopropane to give propylene is thought to proceed via a trimethylene type transition state

Therefore the entropy of activation is positive. The isomerisation of vinyl-allyl ether to give allylacetaldehyde involves a transition state in which internal rotations are reduced to vibrations and results in an increase in order. Such reactions have negative entropies of activation. The frequency factors of some unimolecular reactions are given in table 5.2

TABLE 5.2

Reaction	A/s^{-1}	$\Delta S_{298}^{\ddagger}/\text{J K}^{-1}\ \text{mol}^{-1}$
$CH_3N_2CH_3 \rightarrow C_2H_6 + N_2$	3.5×10^{16}	71
cyclopropane $\rightarrow CH_3CH=CH_2$	1.5×10^{16}	46
Decomposition of dicyclopentadiene	1.0×10^{13}	0
Vinyl allyl ether $\rightarrow$ allylacetaldehyde	5.0×10^{11}	-21
Decomposition of ethylidene diacetate	2×10^{10}	-50

Problems

1. Calculate the rate constant for the reaction

$$H_2 + I_2 \rightarrow 2HI$$

at 556 K and 1 atm pressure assuming the collision diameters are 200 pm for hydrogen and 500 pm for iodine and the activation energy is 170 kJ mol^{-1}.

2. The rate constant for a certain unimolecular reaction is 4.14×10^{-4} s^{-1} at 404 K. Given that the activation energy is 108 kJ mol^{-1} calculate the entropy of activation at this temperature.

3. The unimolecular decomposition of diacetyl has a frequency factor of 8.0×10^{15} s^{-1} at 285°C. Calculate the entropy of activation.

4. Calculate the entropy of activation for the reaction

$$NO + H_2 \rightarrow \tfrac{1}{2}N_2 + H_2O$$

at 827°C given that the rate constant is 145.5 dm^6 mol^{-2} s^{-1} and the activation energy is 184 kJ mol^{-1}.

Further reading

Reference

1. W. F. K. Wynne-Jones and H. Eyring. *J. chem. Phys.*, **3** (1935), 492.

Reviews

K. J. Laidler and J. C. Polanyi. Theories of bimolecular reactions. *Progr. Reaction. Kinetics*, **1** (1963), 41.
B. S. Rabinovitch and M. C. Flowers. Chemical activation. *Q. Rev., chem. Soc.*, **18** (1964), 122.

Books

H. S. Johnson. *Gas-Phase Reaction Theory*, Ronald, New York (1966).
K. J. Laidler. *Chemical Kinetics* (2nd Edition), McGraw-Hill, New York (1965).
G. L. Pratt. *Gas Kinetics*, Wiley, London (1969).

6 THEORY OF UNIMOLECULAR REACTIONS

In a unimolecular reaction a single reactant molecule isomerises or decomposes to give a product or products. In terms of reaction-rate theory, the transition state or activated complex has a configuration similar to the reactant so that the process can be represented by

$$A \rightarrow A^{\ddagger} \rightarrow \text{products}$$

In the 1920s a number of gas-phase decompositions (for example, dinitrogen pentoxide, dimethyl ether, acetone) were found to obey first-order kinetics, and at first were thought to be elementary processes. However, it was soon shown that these reactions were not unimolecular processes, but chain reactions in which the first step was often unimolecular to give free radicals. Many isomerisation processes are unimolecular, for example the isomerisation of cyclopropane to propylene

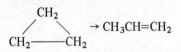

In the early days it was difficult to explain how molecules were activated in a unimolecular process. If activation was by collisions between the molecules, it was assumed that the system would show second-order kinetics. It was thought that the molecules absorbed their energy of activation from the radiation emitted by the vessel walls, but this theory was disproved when the rate constants of unimolecular reactions were found to depend on the volume of the reaction vessel.

6.1 Lindemann theory
In 1922 Lindemann[1] showed that unimolecular reactions do acquire their energy of activation by bimolecular collisions, but that this can give rise to first-order kinetics except at low pressures. His theory was an important development and still forms the basis of all modern theories of unimolecular reactions.

His theory assumed that reactant molecules are activated by collisions with each other, that is by bimolecular collisions. He postulated that there is a time lag between the activation and the reaction of these 'energised molecules' to give products. As a consequence, most of the energised molecules collide with a normal reactant molecule before they can react, lose their excess energy and are deactivated. Provided the rate of deactivation is much greater than the unimolecular decomposition of the energised molecules to give products, the energised molecules are in equilibrium with the normal molecules. This results in a stationary state or steady-state concentration of energised molecules; that is, their concentration remains steady and does not change with time. This so-called *steady-state hypothesis* is discussed on page 82. At high pressures this condition is satisfied and the steady-state concentration of energised molecules is proportional to the concentration of normal molecules. The rate of reaction, given by the rate of conversion of energised molecules into product, is proportional to the concentration of energised molecules, and consequently the concentration of normal molecules. Therefore, at high pressures the reaction is first order.

At low pressures the rate of deactivation decreases as the molecular collision rate decreases, and the rate of conversion of energised molecules to product becomes comparable to their rate of deactivation. Under these conditions the rate of reaction depends on the rate of activation of the energised molecules (a bimolecular process) and the overall kinetics becomes second order.

The mechanism for the reaction can be represented by the following processes:

Activation

$$A + A \xrightarrow{k_1} A^* + A \tag{1}$$

Deactivation

$$A^* + A \xrightarrow{k_{-1}} A + A \tag{-1}$$

Unimolecular decomposition

$$A^* \xrightarrow{k_2} products \tag{2}$$

where A and A* represent a normal and energised molecule, respectively.

Since A* molecules are formed by reaction (1) and lost by reactions (-1) and (2), their rate of formation is given by the rate of reaction (1) minus the sum of the rates of reactions (-1) and (2); that is

$$\frac{d[A^*]}{dt} = k_1[A]^2 - k_{-1}[A^*][A] - k_2[A] \tag{6.1}$$

Assuming that a steady-state concentration of energised molecules exists, that is their concentration does not change with time, this expression can be equated to zero giving

$$\frac{d[A^*]}{dt} = 0 \qquad (6.2)$$

Combining equations 6.1 and 6.2 gives

$$[A^*] = \frac{k_1[A]^2}{k_{-1}[A] + k_2} \qquad (6.3)$$

The rate of reaction v (that is, the rate of formation of product) is given by the rate of reaction (2)

$$v = k_2[A^*] = \frac{k_1 k_2 [A]^2}{k_{-1}[A] + k_2} \qquad (6.4)$$

At high pressures where the rate of deactivation is much greater than the rate of conversion to products, that is $k_{-1}[A][A^*] \gg k_2[A^*]$, equation 6.4 becomes

$$v = \frac{k_1 k_2 [A]}{k_{-1}} = k_\infty [A] \qquad (6.5)$$

Therefore the reaction is first order and the limiting or high-pressure first-order rate constant k_∞ is equal to $k_1 k_2 / k_{-1}$.

At low pressures, the rate of deactivation becomes less than the rate of conversion to product, that is $k_{-1}[A][A^*] \ll k_2[A^*]$, so that equation 6.4 becomes

$$v = k_1[A]^2 \qquad (6.6)$$

Therefore at low pressures the reaction is second order.

It has been shown that Lindemann theory predicts a change in order when the pressure is increased or decreased.

Let the rate of reaction at any pressure be given by

$$v = k'[A] \qquad (6.7)$$

where k' is a rate coefficient that varies with pressure. From equation 6.4 it is seen that k' is given by

$$k' = \frac{k_1 k_2 [A]}{k_{-1}[A] + k_2}$$

or

$$k' = \frac{k_\infty}{1 + k_2[A]/k_{-1}} \qquad (6.8)$$

Equation 6.8 predicts that a plot of k' against $[A]$ will be as shown in figure 6.1 and that k' will have a limiting value of k_∞ at high pressures, but falls off to zero at low pressures.

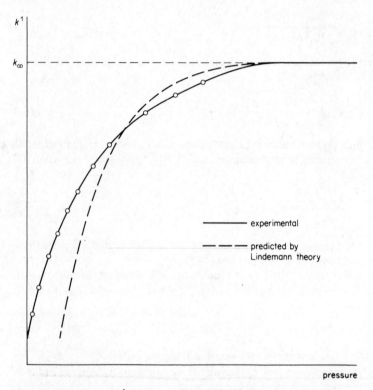

Figure 6.1 *Plot of k' against pressure for a unimolecular reaction*

Experimental rate data for unimolecular reactions is in qualitative agreement with Lindemann theory. If the half-life for the decomposition is plotted against pressure, it is found to be a constant at high pressure but increases at low pressures as the kinetics change from first order to second order. However it has been shown that the fall off in the rate constant occurs at higher pressures than that predicted by Lindemann as illustrated in figure 6.1.

Alternatively Lindemann theory can be tested by inverting equation 6.8 giving

$$\frac{1}{k'} = \frac{k_{-1}}{k_1 k_2} + \frac{1}{k_1[A]} \tag{6.9}$$

A plot of $1/k'$ against $1/[A]$ should therefore be a straight line of slope $1/k_1$ as shown in figure 6.2 for the 1,1-dimethylcyclopropane isomerisation studied by Flowers and Frey.[2] Again it is found that deviations from linearity occur at high pressures. Modern theories of unimolecular reactions develop Lindemann theory and seek to explain these deviations.

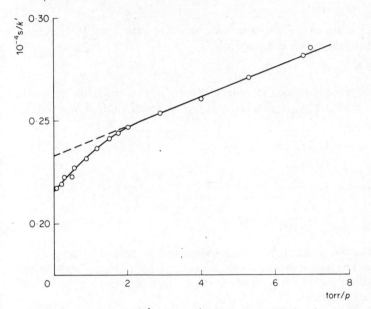

Figure 6.2 *A plot of $1/k'$ against $1/p$ for 1,1-dimethylcyclopropane isomerisation*

Further evidence for the basic soundness of Lindemann theory can be obtained when the reaction is carried out with a constant reactant pressure and the total pressure varied by the addition of an inert gas M such as nitrogen, argon or xenon. The mechanism is now represented by

$$A + M \xrightarrow{k_1} A^* + M$$
$$A^* + M \xrightarrow{k_{-1}} A + M$$
$$A^* \xrightarrow{k_2} \text{products}$$

Steady-state treatment of this mechanism leads to

$$v = \frac{k_1 k_2 [A][M]}{k_{-1}[M] + k_2}$$

At high pressures $k_{-1}[M] \gg k_2$ giving

$$v = \frac{k_1 k_2}{k_{-1}} [A] = k_\infty [A]$$

which is identical to the high-pressure rate equation 6.5.

At low pressures $k_{-1}[M] \ll k_2$ giving

$$v = k_1[A][M]$$

that is, the reaction is first order with respect to both A and M. Therefore the rate of reaction is expressed by

$$v = k'[A]$$

where k' is the observed first-order rate coefficient, which has limiting values at high pressures of $k_1 k_2/k_{-1}$ and at low pressures of $k_1[M]$ or in general is given by

$$k' = \frac{k_1 k_2 [M]}{k_{-1}[M] + k_2}$$

Inverting

$$\frac{1}{k'} = \frac{k_{-1}}{k_1 k_2} + \frac{1}{k_1[M]}$$

A plot of $1/k'$ against $1/[M]$ is linear, showing that the added inert gas can replace the reactant molecule as an activator and deactivator. These experiments illustrate the basic soundness of the Lindemann mechanism for unimolecular reactions.

6.2   Hinshelwood theory

In 1927 Hinshelwood[3] postulated that the rate of energisation of a molecule depends on the number of vibrational degrees of freedom in the molecule. A molecule with a large number of vibrational degrees of freedom has a much greater probability of acquiring the energy needed for activation, since this energy can be distributed among all these degrees of freedom.

For a molecule with one degree of freedom, the rate constant for the energisation process (1) is given by

$$k_1 = Z_1 \exp{(-E^{\ddagger}/RT)} \qquad (6.10)$$

where Z_1 is the bimolecular collision number and $E^{\ddagger}$ is the energy acquired. But for a molecule with s degrees of vibrational freedom

$$k_1 = \frac{Z_1}{(s-1)!} \left(\frac{E^{\ddagger}}{RT}\right)^{s-1} \exp\left(-E^{\ddagger}/RT\right) \tag{6.11}$$

and this results in a much higher value of k_1.

Lindemann theory predicted that first-order behaviour would be maintained down to much lower pressures than that observed experimentally because the values of k_1 were calculated from equation 6.10 instead of equation 6.11.

Example 6.1

Calculate the frequency factor for a reaction at 300 K with an activation energy of 200 kJ mol^{-1} and $s = 6$, assuming the collision number $Z_1 = 10^{12}$ dm^3 mol^{-1} s^{-1}.

The pre-exponential term according to Hinshelwood is given by

$$\frac{Z_1}{(s-1)!} \left(\frac{E^{\ddagger}}{RT}\right)^{s-1} = \frac{10^{12}}{5!} \left(\frac{200 \times 10^3}{8.31 \times 300}\right)^5 \text{dm}^3 \text{ mol}^{-1} \text{s}^{-1}$$

$$= 2.8 \times 10^{19} \text{ dm}^3 \text{ mol}^{-1} \text{s}^{-1}$$

This gives a frequency factor and consequently a rate constant, which is 10^7 times greater than that predicted by collision theory since $Z_1 = 10^{12}$ dm^3 mol^{-1} s^{-1}. This corresponds to a much higher rate of activation and, as a result, first-order dependence falls off at much lower pressures than predicted by Lindemann.

An unsatisfactory feature of Hinshelwood's theory is that the values of s were determined by trial and error, and in most cases, the best experimental fit was obtained with s corresponding to about half the total number of vibrational degrees of freedom. It is possible that only a certain number of the total degrees of freedom are involved in the formation of the activated complex.

In the light of later theories the basic mechanism for a unimolecular reaction is best represented by the following modification of Lindemann's mechanism

$$A + A \xrightarrow{k_1} A^* + A \tag{1}$$

$$A + A^* \xrightarrow{k_{-1}} A + A \tag{-1}$$

$$A^* \xrightarrow{k_{2a}} A^{\dagger} \tag{2a}$$

$$A^{\dagger} \xrightarrow{k_{2b}} \text{products} \tag{2b}$$

where $A^\dagger$ is an activated molecule. The energised molecule A* has sufficient energy to be chemically activated without further acquisition of energy. It undergoes vibrational energy changes and is activated. When the energy becomes localised in a particular bond or bonds, it is converted into product. Modern theories predict that molecules can be energised more readily than predicted by Lindemann but that the time-lag between energisation and activation or reaction is often relatively long.

6.3   RRK and Slater theory

A full treatment of the modern theories of Rice, Ramsperger[4] and Kassel[5] (known as RRK theory) and Slater is beyond the scope of this book. RRK theory seeks to explain why plots of the type shown in figure 6.2 are not linear. It proposes that a molecule is activated when the critical amount of energy is concentrated in one particular bond. It is assumed that the energy redistributes itself freely between the normal vibrational modes during each vibration of the molecule. The rate constant k_{2b} is therefore of the same magnitude as the average vibration frequency of the molecule.

On the other hand Slater's theory proposes that energy is not free to flow within the molecule. Slater suggests that reaction occurs when a

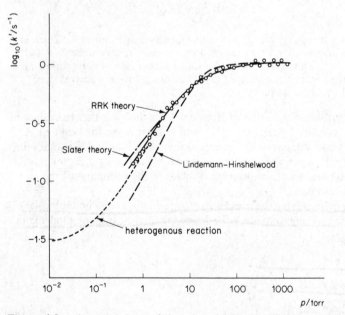

Figure 6.3   *A comparison of theoretical fall-off curves for the isomerisation of cyclopropane at* $500°C$

'critical coordinate' in the molecule, usually a bond length, becomes extended by a critical amount. In a complex molecule each of the vibrational modes vibrate at different frequencies, and this critical extension of a bond occurs when two stretching modes are in phase.

Figure 6.3 shows a theoretical treatment of the experimental results[6] for the isomerisation of cyclopropane in terms of the above theories.

Problems

1. Hinshelwood and Ashley (*Proc. R. Soc.*, **A115** (1927), 215) obtained the following overall rate constants k' for the decomposition of dimethyl ether at 773 K

Initial conc./mmol dm^{-3}	1.20	1.89	3.55	5.42	8.18	13.21	18.57
$10^4 k'/s^{-1}$		2.48	3.26	4.61	5.54	6.29	6.90	7.45

Use Lindemann theory to calculate the limiting rate constant k_∞ and the rate constant of the collisional activation step k_1.

2. Calculate the rate constant k_1 for the collisional activation step of a unimolecular reaction, if the activation energy is 167 kJ mol^{-1} and $s = 12$.

3. Determine the ratio of the rate constant calculated by simple collision theory to that calculated using the Hinshelwood expression for a unimolecular reaction with an activation energy of 80 kJ mol^{-1} and $s = 8$.

Further reading

References

1. F. A. Lindemann. *Trans. Faraday Soc.*, **17** (1922), 598.
2. M. C. Flowers and H. M. Frey. *J. chem. Soc.* (1962), 1160.
3. C. N. Hinshelwood. *Proc. R. Soc.*, **A113** (1927), 230.
4. O. K. Rice and H. C. Ramsperger. *J. Am. chem. Soc.*, **49** (1927), 1617; *J. Am. chem. Soc.*, **50** (1928), 617.
5. L. S. Kassel. *J. chem. Phys.*, **32** (1928), 225.
6. H. O. Pritchard, R. G. Sowden and A. F. Trotman-Dickenson. *Proc. R. Soc.*, **A218** (1953), 224.

Review

B. G. Gowenlock. Arrhenius factors in unimolecular reactions. *Q. Rev. chem. Soc.*, **14** (1960), 133.

Books

P. J. Robinson and K. A. Holbrook. *Unimolecular Reactions*, Wiley, London (1971).
N. B. Slater. *Theory of Unimolecular Reactions*, Methuen, London (1959).

7 ATOMIC AND FREE-RADICAL PROCESSES

In the early days of chemical kinetics it was assumed that all reactions took place in a single step according to the stoichiometric equation. It is now clear that the majority of chemical processes proceed via a number of steps, so that most reactions are essentially complex. It has been shown that in many reactions reactive intermediates such as atoms and free radicals play an important role.

For kinetic purposes a free radical can be defined as an atomic or molecular species which contains one or more unpaired electrons. Monoradicals contain one unpaired electron while diradicals such as ground-state oxygen atoms contain two unpaired electrons. As a consequence radicals are paramagnetic and are chemically reactive. Molecules such as nitric oxide, oxygen and 2,2-diphenyl-1-picrylhydrazyl, which contain unpaired electrons, can also be regarded as free radicals by this definition.

The kinetics of reactions involving free radicals are often complex, but experimental rate data has proved to be a very fruitful tool in the elucidation of the mechanisms of such reactions. The aim of the kineticist is to postulate a reaction mechanism that is in qualitative and quantitative agreement with all the experimental rate data for that reaction. The more reliable the rate data for the elementary steps in the proposed reaction scheme, the greater is the measure of confidence in the proposed reaction mechanism.

7.1 Types of complex reaction
Complex reactions can be classified into the following groups: non-chain processes, linear-chain processes and branched-chain processes.

7.1.1 Non-chain processes
In a non-chain complex reaction, an active centre such as a free radical or molecule is formed. This reacts to give an intermediate and hence a product. In no way is it possible for the intermediate to be regenerated. An example of a non-chain complex reaction is the iodination of acetone in acid solution, which proceeds as follows

$$CH_3COCH_3 \xrightarrow{\text{acid}} CH_3C{=}CH_2$$
$$\underset{OH}{|}$$

$$CH_3C=CH_2 + I_2 \longrightarrow CH_3CICH_2I$$
$$\quad\;\; |\qquad\qquad\qquad\qquad\quad |$$
$$\quad OH \qquad\qquad\qquad\qquad OH$$

$$CH_3CICH_2I \longrightarrow HI + CH_3COCH_2I$$
$$\quad\;\; |$$
$$\quad OH$$

7.1.2 Linear chain processes

A chain process is one that proceeds by a succession of elementary processes as follows.

(i) Chain initiation

The reaction is initiated when the weakest bond in the reactant or one of the reactants is broken to produce a free radical, which acts as a chain carrier.

(ii) Chain propagation

The free radical attacks the reactant to produce a product molecule and another reactive species. This new free radical reacts further to regenerate the original free radical, which once again attacks the reactant molecule. In this way the product and the chain carrier are formed continuously. These processes are termed propagation reactions.

(iii) Chain termination

In addition the free radicals are removed from the reaction system by recombination or disproportionation. In this way the chain carriers are destroyed and the chains terminate.

These steps are characteristic of any chain reaction.

7.1.3 Branched chain processes

In some reactions, particularly gas-phase hydrocarbon oxidations, there is a continuous build up of free radicals in the system. This usually arises when in one or more steps one free radical reacts to produce two or more free radicals. In the hydrogen–oxygen reaction, two such steps are

$$H\cdot + O_2 \rightarrow OH\cdot + O\colon$$

$$O\colon + H_2 \rightarrow OH\cdot + H\cdot$$

They occur because molecular oxygen and ground-state atomic oxygen are biradical species. In these reactions the concentration of free radicals increases very rapidly, as illustrated in figure 7.1, and this is known as chain branching. The reaction rate increases very rapidly and soon becomes (theoretically) infinite, causing an explosion to occur.

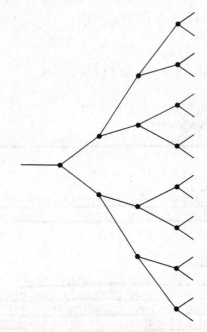

Figure 7.1 *Illustration of the rapid growth in the number of free radicals via chain branching.*

7.1.4 Steady- or stationary-state approximation

In a linear chain process, steady-state conditions soon prevail. After a short induction time when the concentration of free radicals rises, their concentration becomes steady or stationary, and does not change with time until the reactants are consumed. This means that the rate at which the free radicals are formed is equal to the rate at which they are destroyed; that is

$$\frac{d[\text{radical}]}{dt} = 0 \qquad (7.1)$$

It is customary to assume that all the free radicals in the reaction system attain a steady state very quickly. This approximation assists greatly in the derivation of a rate equation for a chain process. Without this it would be necessary to solve a number of differential equations. This would be a tedious task without the help of a computer.

7.2 Hydrogen–bromine reaction

The reaction between hydrogen and bromine gas at temperatures between $200°C$ and $300°C$ was studied by Bodenstein and Lind[1] in 1906. This was subsequently shown to be a linear chain reaction. By contrast the $H_2 + I_2$ reaction was thought to be a simple bimolecular reaction. The $H_2 + Br_2$ reaction is a good example of a chain reaction, and is the classical example quoted in most physical chemistry textbooks. It can be shown not only that the proposed mechanism is consistent with the experimental rate data, but that other possible elementary steps are not important in this reaction.

Bodenstein and Lind's experimental results gave the rate equation as

$$\frac{d[HBr]}{dt} = \frac{k[H_2][Br_2]^{1/2}}{1 + k'[HBr]/[Br_2]} \tag{7.2}$$

where k' was equal to about 10 and found to be independent of temperature. The following five-step mechanism was later proposed to explain their experimental result.

$Br_2 \xrightarrow{k_1} Br\cdot + Br\cdot$	Chain initiation	(1)
$Br\cdot + H_2 \xrightarrow{k_2} HBr + H\cdot$	Chain propagation	(2)
$H\cdot + Br_2 \xrightarrow{k_3} HBr + Br\cdot$	Chain propagation	(3)
$H\cdot + HBr \xrightarrow{k_{-2}} H_2 + Br\cdot$	Chain inhibition	(−2)
$Br\cdot + Br\cdot \xrightarrow{k_{-1}} Br_2$	Chain termination	(−1)

This has all the characteristics of a linear-chain process. Step (1) is the initiation reaction, steps (2) and (3) propagate the chain, and step (−1) is the termination reaction. The unusual step is reaction (−2) where the product is attacked by a free radical. The result is a rare example of a reaction whose rate is influenced by the concentration of the product. The reactive intermediates or chain carriers are hydrogen and bromine atoms, which are continuously generated by the propagation steps.

In order to show that the proposed mechanism is consistent with the experimental results, it is necessary to derive the rate equation. The following procedure is a good guide to a general approach to any such derivation:

(1) Express the required rate in terms of the rates of the elementary steps involved.
(2) Apply the steady-state approximation to all the free radicals in the reaction.
(3) By manipulation of the algebraic equations, express the free-radical concentrations in terms of reactant concentrations only.
(4) Hence eliminate free-radical concentrations from the rate equation, which is then expressed in the simplest possible mathematical form.

(1) The required rate is the rate of formation of hydrogen bromide; that is

$$\frac{d[HBr]}{dt} = k_2[Br][H_2] + k_3[H\cdot][Br_2] - k_{-2}[H\cdot][HBr] \tag{7.3}$$

(2) Application of the steady-state approximation to [Br] and [H] gives

$$\frac{d[Br\cdot]}{dt} = 2k_1[Br_2] - k_2[Br\cdot][H_2] + k_3[H\cdot][Br_2] + k_{-2}[H\cdot][HBr]$$
$$-2k_{-1}[Br\cdot]^2 = 0 \tag{7.4}$$

and

$$\frac{d[H\cdot]}{dt} = k_2[Br\cdot][H_2] - k_3[H\cdot][Br_2] - k_{-2}[H\cdot][HBr] = 0 \tag{7.5}$$

(3) Addition of equations 7.4 and 7.5 gives

$$2k_1[Br_2] - 2k_{-1}[Br\cdot]^2 = 0$$

so that

$$[Br\cdot] = \left(\frac{k_1}{k_{-1}}\right)^{1/2} [Br]^{1/2} \tag{7.6}$$

From equation 7.5

$$[H\cdot] = \frac{k_2[H_2][Br]}{k_3[Br_2] + k_{-2}[HBr]} \tag{7.7}$$

Substitution of equation 7.6 in equation 7.7 gives

$$[H\cdot] = \frac{k_2(k_1/k_{-1})^{1/2}[H_2][Br_2]^{1/2}}{k_3[Br_2] + k_{-2}[HBr]} \tag{7.8}$$

(4) Equation 7.3 can be simplified by adding it to equation 7.5, giving

$$\frac{d[HBr]}{dt} = 2k_3[H\cdot][Br_2] \tag{7.9}$$

Substitution of equation 7.8 in equation 7.9 gives

$$\frac{d[HBr]}{dt} = \frac{2k_2(k_1/k_{-1})^{1/2}[H_2][Br_2]^{3/2}}{k_3[Br_2] + k_{-2}[HBr]}$$

Dividing throughout by $k_3[Br_2]$ gives

$$\frac{d[HBr]}{dt} = \frac{2k_2(k_1/k_{-1})^{1/2}[H_2][Br_2]^{1/2}}{1 + k_{-2}[HBr]/k_3[Br_2]} \tag{7.10}$$

It can be seen that equation 7.10 is equivalent to equation 7.2 when

$$k = 2k_2(k_1/k_{-1})^{1/2}$$

and

$$k' = k_{-2}/k_3$$

It can also be shown that other possible steps are not important in this reaction.

The initiation step

$$H_2 \rightarrow H\cdot + H\cdot$$

and the alternative inhibition step

$$Br\cdot + HBr \rightarrow H\cdot + Br_2$$

are too slow to be involved. The concentration of H atoms is about 10^{-6} times that of the bromine atom concentration, so that termination steps involving H atoms can be ignored. The excellent agreement between equation 7.10 and the experimental rate equation also indicates that the other processes are slow relative to reactions (1), (2), (3), (−2) and (−1).

7.3 Rice–Herzfeld mechanisms

7.3.1 Paneth's lead-mirror experiment

One of the first techniques used to show the importance of free radicals in the decomposition of organic compounds in the gas phase was developed by Paneth.[2] He passed a stream of hydrogen through a vessel containing lead tetramethyl. A stream of hydrogen saturated with lead tetramethyl passed down a reaction tube as shown in figure 7.2.

A furnace was placed at position A and a lead mirror was deposited on the cool section of the tube just beyond the furnace. It was proposed that

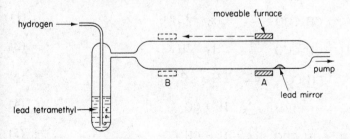

Figure 7.2 *Paneth's apparatus for the removal of lead mirrors by methyl radicals*

the decomposition of the vapour produced a deposit of lead and free methyl radicals, which were pumped away.

$$Pb(CH_3)_4 \rightarrow Pb\downarrow + 4CH_3\cdot \tag{1}$$

The furnace was then moved upstream to position B some 20 cm from A. It was found that not only was a new lead mirror formed at B, but that the original lead mirror at A slowly disappeared. The rate of its disappearance was found to decrease as the distance AB increased. From the above, it is apparent that free methyl radicals formed in reaction (1) attack the original lead mirror and form volatile lead tetramethyl, which is pumped away

$$Pb + 4CH_3\cdot \rightarrow Pb(CH_3)_4 \tag{2}$$

As the distance AB is increased, more and more methyl radicals recombine to form ethane by

$$CH_3\cdot + CH_3\cdot \rightarrow C_2H_6 \tag{3}$$

and the rate of methyl attack on the lead by reaction (2) decreases.

7.3.2 Thermal decomposition of acetaldehyde

In the early days of kinetics many organic pyrolyses were found to be first or second order and were assumed to be molecular processes. It was then shown that free radicals were important chain carriers in these reactions. Rice and Herzfeld[3] were among the first workers to suggest chain mechanisms for these pyrolyses.

One of the simplest examples of a Rice–Herzfeld mechanism is provided by the thermal decomposition of acetaldehyde. A simplified mechanism for this reaction was given on page 3, but a more detailed mechanism is given here.

$$CH_3CHO \xrightarrow{k_1} CH_3\cdot + CHO\cdot \tag{1}$$

$$CH_3\cdot + CH_3CHO \xrightarrow{k_2} CH_4 + CH_3CO\cdot \tag{2}$$

$$CH_3CO\cdot \xrightarrow{k_3} CH_3\cdot + CO \tag{3}$$

$$CHO\cdot \xrightarrow{k_4} H\cdot + CO \tag{4}$$

$$H\cdot + CH_3CHO \xrightarrow{k_5} H_2 + CH_3CO\cdot \tag{5}$$

$$CH_3\cdot + CH_3\cdot \xrightarrow{k_6} C_2H_6 \tag{6}$$

In this reaction scheme the initiation step produces methyl and formyl radicals. Methyl radicals react to produce methane and acetyl radicals. Formyl and acetyl radicals decompose in unimolecular reactions to give carbon monoxide and radicals. The major termination step produces ethane.

The main products of this reaction are CH_4 and CO, with H_2 and C_2H_6 produced as minor products. The experimental rate equation was found to be

$$-\frac{d[CH_3CHO]}{dt} = k_r[CH_3CHO]^{3/2} \tag{7.11}$$

and it is necessary to show that the above mechanism gives a rate expression of this form.

From the above mechanism, the rate of decomposition of acetaldehyde is given by

$$-\frac{d[CH_3CHO]}{dt} = k_1[CH_3CHO] + k_2[CH_3 \cdot][CH_3CHO]$$
$$+ k_5[H \cdot][CH_3CHO] \tag{7.12}$$

Applying the steady-state approximation to all the free radicals gives

$$\frac{d[CH_3 \cdot]}{dt} = k_1[CH_3CHO] - k_2[CH_3 \cdot][CH_3CHO] + k_3[CH_3CO \cdot]$$
$$- 2k_6[CH_3 \cdot]^2 = 0 \tag{7.13}$$

$$\frac{d[CHO \cdot]}{dt} = k_1[CH_3CHO] - k_4[CHO \cdot] = 0 \tag{7.14}$$

$$\frac{d[CH_3CO \cdot]}{dt} = k_2[CH_3 \cdot][CH_3CHO] - k_3[CH_3CO \cdot]$$
$$+ k_5[H \cdot][CH_3CHO] = 0 \tag{7.15}$$

$$\frac{d[H \cdot]}{dt} = k_4[CHO \cdot] - k_5[H \cdot][CH_3CHO] = 0 \tag{7.16}$$

Addition of equations 7.14 and 7.16 gives

$$[H \cdot] = k_1/k_5 \tag{7.17}$$

Similarly, addition of equations 7.13 and 7.15 gives

$$k_1[CH_3CHO] - 2k_6[CH_3 \cdot]^2 + k_5[H \cdot][CH_3CHO] = 0 \tag{7.18}$$

Substitution of equation 7.16 into 7.17 gives

$$k_1[CH_3CHO] = k_6[CH_3\cdot]^2$$

that is

$$[CH_3\cdot] = (k_1/k_6)^{1/2}[CH_3CHO]^{1/2} \tag{7.19}$$

Substitution of equations 7.17 and 7.19 into equation 7.12 gives

$$-\frac{d[CH_3CHO]}{dt} = 2k_1[CH_3CHO] + k_2(k_1/k_6)^{1/2}[CH_3CHO]^{3/2} \tag{7.20}$$

Assuming that the initiation and termination steps are slow relative to the propagation steps, the first term in equation 7.20 can be neglected and the rate equation becomes

$$-\frac{d[CH_3CHO]}{dt} = k_2 \left(\frac{k_1}{k_6}\right)^{1/2} [CH_3CHO]^{3/2} \tag{7.21}$$

which is consistent with the experimental rate equation 7.11.

Recent work by Laidler and Liu[4] has suggested that other processes play a part in the reaction and give rise to minor products such as acetone and propionaldehyde. These arise from additional propagation processes such as

$$CH_3\cdot + CH_3CHO \rightarrow CH_4 + CH_2CHO\cdot$$

and

$$CH_3\cdot + CH_3CHO \rightarrow H\cdot + CH_3COCH_3$$

and the termination process

$$CH_3\cdot + CH_2CHO\cdot \rightarrow CH_3CH_2CHO$$

which occur to a very small extent.

7.3.3 Activation energy
One feature of a Rice–Herzfeld type decomposition is that the overall activation energy is usually very much less than the energy needed to cleave the C—C bond in the initiation process. This can be illustrated by the acetaldehyde pyrolysis.

The rate constant k_r for the reaction is given by

$$k_r = k_2 \left(\frac{k_1}{k_6}\right)^{1/2}$$

In terms of the frequency factor and activation energy of the individual steps

$$k_r = A_2 \exp\left(-E_2^{\ddagger}/RT\right) \frac{A_1 \exp\left(-E_1^{\ddagger}/RT\right)^{1/2}}{A_6 \exp\left(-E_6^{\ddagger}/RT\right)^{1/2}}$$

$$= A_2 \left(\frac{A_1}{A_6}\right)^{1/2} \exp\left(\frac{-[E_2^{\ddagger} + \frac{1}{2}(E_1^{\ddagger} - E_6^{\ddagger})]}{RT}\right)$$

The overall activation energy is, therefore, given by

$$E^{\ddagger} = E_2^{\ddagger} + \tfrac{1}{2}(E_1^{\ddagger} - E_6^{\ddagger})$$

Since the activation energy for the initiation step is 332 kJ mol^{-1}, and the activation energy for the termination step is zero, $E^{\ddagger}$ can be calculated if $E_2^{\ddagger}$ is known. From the corresponding photodecomposition of acetaldehyde a value of 32 kJ mol^{-1} for $E_2^{\ddagger}$ is obtained. Substitution of these figures gives

$$E^{\ddagger} = 32 + \tfrac{1}{2}(332 - 0) \text{ kJ mol}^{-1}$$
$$= 198 \text{ kJ mol}^{-1}$$

This is in excellent agreement with the experimental value for the activation energy of 193 kJ mol^{-1} and is seen to be much less than the energy ($\geqslant$332 kJ mol^{-1}) needed to break the original C—C bond.

7.4   Addition polymerisation

Addition polymerisation processes provide excellent examples of linear free-radical chain reactions. Once the polymerisation has been initiated by a free radical, the original monomer molecule continues to grow in length forming a large polymeric radical. These radicals eventually recombine or disproportionate to give the polymer product.

Addition polymerisation is initiated by free radicals generated when a suitable initiator molecule is decomposed thermally or photochemically. Benzoyl peroxide decomposes at 70°-100°C in solution and is often used as an initiator

$$C_6H_5CO_2—O_2CC_6H_5 \rightarrow 2C_6H_5CO_2\cdot \rightarrow 2C_6H_5\cdot + 2CO_2$$

Acetone is readily decomposed photochemically

$$CH_3COCH_3 \xrightarrow{h\nu} 2CH_3\cdot + CO$$

The subsequent propagation and termination processes can be illustrated by reference to an olefinic monomer $CH_2{=}CHX$, where X is H for ethylene,

Cl for vinyl chloride and C_6H_5 for styrene. If R· is the free radical formed in the initiation process, propagation proceeds as follows

$$R\cdot + CH_2{=}CHX \rightarrow RCH_2\dot{C}HX$$

$$RCH_2\dot{C}HX + CH_2{=}CHX \rightarrow RCH_2CHXCH_2\dot{C}HX$$

$$R(CH_2CHX)_{n-1}CH_2\dot{C}HX + CH_2{=}CHX \rightarrow R(CH_2CHX)_nCH_2\dot{C}HX$$

These radicals continue to grow until they undergo termination by either:

(i) *recombination*, in which two unpaired electrons pair to form a single bond

$$R(CH_2CHX)_nCH_2\dot{C}HX + \dot{C}HXCH_2(CHXCH_2)_nR \rightarrow$$

$$R(CH_2CHX)_nCH_2CHXCHXCH_2(CHXCH_2)_nR$$

or (ii) *disproportionation*, in which there is a transfer of a hydrogen atom to form both a saturated and an unsaturated polymer molecule

$$R(CH_2CHX)_nCH_2\dot{C}HX + \dot{C}HXCH_2(CHXCH_2)_nR \rightarrow$$

$$R(CH_2CHX)_nCH_2CH_2X + CHX{=}CH(CHXCH_2)_nR$$

The kinetics of addition polymerisation reactions can be derived from the following general mechanism

$$
\begin{array}{lll}
I & \xrightarrow{k_i} \alpha R_1\cdot & \text{Initiation} \\
R_1\cdot + M & \xrightarrow{k_p} R_2\cdot & \\
R_2\cdot + M & \xrightarrow{k_p} \cdot R_3\cdot & \\
R_3\cdot + M & \xrightarrow{k_p} R_4\cdot & \text{Propagation} \\
R_{n-1} + M & \xrightarrow{k_p} R_n\cdot & \\
R_n\cdot + R_m\cdot & \xrightarrow{k_t} P_{n+m} & \text{Termination}
\end{array}
$$

where I represents an initiator molecule, M a monomer molecule, P a polymer molecule, α is the number of free radicals obtained from each molecule of initiator and $R_1\cdot$, $R_2\cdot$, $R_3\cdot$, etc., are free radicals. It is found that the rate constant k_p for all the propagation processes is the same and that similarly k_t can be assumed to be the rate constant for all the termination processes. The rate of initiation $v_i = k_i[I]$, where k_i is the rate constant for initiation.

Application of the steady-state approximation to the free radicals in the system gives

$$\frac{d[R_1\cdot]}{dt} = v_i - k_p[R_1\cdot][M] - k_t[R_1\cdot]([R_1\cdot] + [R_2\cdot] + \ldots) = 0$$

where $k_t[R_1\cdot]^2$, $k_t[R_1\cdot][R_2\cdot]$, etc., are the rates of the termination processes $R_1\cdot + R_1\cdot$, $R_1\cdot + R_2\cdot$, etc., respectively.

Therefore

$$\frac{d[R_1\cdot]}{dt} = v_i - k_p[R_1\cdot][M] - k_t[R_1\cdot] \sum_{n=1}^{\infty} [R_n\cdot] = 0$$

Also

$$\frac{d[R_2\cdot]}{dt} = k_p[R_1\cdot][M] - k_p[R_2\cdot][M] - k_t[R_2\cdot] \sum_{n=1}^{\infty} [R_n\cdot] = 0$$

The radical $R_n\cdot$ is formed by a propagation process, but is removed only by termination processes, therefore

$$\frac{d[R_n\cdot]}{dt} = k_p[R_{n-1}][M] - k_t[R_n\cdot] \sum_{n=1}^{\infty} [R_n\cdot] = 0$$

Summing the steady-state equations, all the k_p terms cancel out giving

$$v_i - k_t \left(\sum_{n=1}^{\infty} [R_n\cdot] \right)^2 = 0$$

The condition for steady-state polymerisation is, therefore, that the rate of initiation is equal to the sum of all the termination rates; that is

$$\left(\sum_{n=1}^{\infty} [R_n\cdot] \right)^2 = \frac{v_i}{k_t}$$

or

$$\sum_{n=1}^{\infty} [R_n\cdot] = \left(\frac{v_i}{k_t} \right)^{1/2}$$

The rate of reaction as measured by the rate of disappearance of monomer is given by

$$-\frac{d[M]}{dt} = k_p[M] \sum_{n=1}^{\infty} [R_n\cdot]$$

or

$$= k_p \left(\frac{v_i}{k_t} \right)^{1/2} [M] \tag{7.22}$$

Equation 7.22 is the general expression for the rate of an addition polymerisation.

For any polymerisation process, the initial concentration of monomer is known and techniques are available for the measurement of the rate of initiation. It is usual to add a known concentration of free radical scavenger or inhibitor such as ferric chloride or diphenylpicrylhydrazyl (DPPH) to the solution. This removes the free radicals as they are formed by the initiation process so that the rate of disappearance of scavenger (often measured spectrophotometrically) is equal to the rate of production of free radicals. Alternatively the concentration of initiator is measured by a sampling method after a given time interval.

Therefore, provided the rate of polymerisation is measured (often by a dilatometer) and v_i is determined by one of the methods described above, a value of $k_p/k_t^{1/2}$ can be determined. This is a characteristic constant for any addition polymerisation.

7.5   Gas-phase autoxidation reactions

The reaction of molecular oxygen with another substance is known as an autoxidation. When the reaction is in the gas phase, a branching chain process is likely to occur. The reactivity of molecular oxygen is not surprising since it is a biradical having two unpaired electrons. As a consequence it undergoes reactions in which one radical produces two radicals. In the hydrogen–oxygen reaction, molecular oxygen reacts with hydrogen atoms to give two reactive species, hydroxyl radicals and oxygen atoms

$$H\cdot + O_2 \rightarrow OH\cdot + O\colon$$

Ground-state oxygen atoms are also biradicals and with molecular hydrogen give hydroxyl radicals and hydrogen atoms

$$O\colon + H_2 \rightarrow OH\cdot + H\cdot$$

Both these processes are chain-branching reactions and result in a rapid increase in the number of free radicals. In such systems steady-state conditions do not hold and the reaction rate increases rapidly as the number of free radicals increases. Under non-stationary conditions the rate of reaction becomes infinite and explosion occurs.

Explosions caused by chain branching therefore arise when the concentration of free radicals in the system increases rapidly. On the other hand thermal explosions occur when the rate of reaction increases due to a sharp rise in temperature. If the heat given out by an exothermic reaction is not dissipated sufficiently rapidly, the temperature rises. Since the reaction rate increases exponentially with temperature, a thermal explosion can soon occur.

7.5.1   Hydrogen–oxygen reaction

The reaction between hydrogen and oxygen proceeds at temperatures between $450°C$ and $600°C$ according to the stoichiometric equation

$$2H_2 + O_2 \rightarrow 2H_2O$$

It is a classical example of a branching chain reaction, which has been studied extensively for many years. The rate is found to depend on the total pressure in a manner that is characteristic of all chain-branching reactions.

Consider the above reaction at 550°C. The variation of rate with total pressure is shown in figure 7.3. At low pressures the rate varies linearly with total pressure as expected in a normal non-branching process. At pressures· above about 150 torr and below about 250 torr, a similar dependence is observed. But at pressures between about 50 torr and 150 torr explosion occurs. Therefore, explosion limits known as the first, second and third explosion limits, respectively, occur as shown.

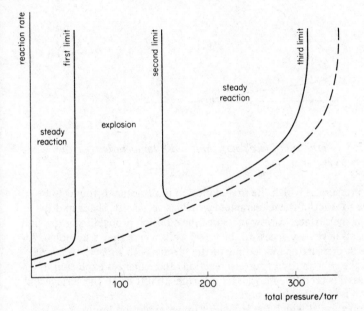

Figure 7.3   *Variation of rate with total pressure for the hydrogen–oxygen reaction.*

The explosion limits are temperature dependent as illustrated by figure 7.4. Below 400°C the reaction proceeds at a steady rate without explosion for a wide range of total pressure. At 500°C the pressure range over which the system is explosive is smaller, since the second explosion limit occurs at a lower pressure. Similarly at this temperature the third explosion limit occurs at higher pressures than at 550°C. At temperatures greater than 600°C the reaction is stable at low pressures but explodes at all other pressures.

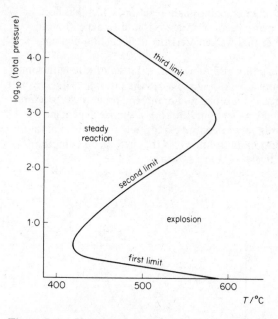

Figure 7.4   *Variation of explosion limits with temperature for the hydrogen–oxygen reaction*

The pressure at which the first explosion limit occurs is found to be sensitive to reaction vessel parameters such as vessel size, shape and the nature of the surface. At low pressures the collision probability is low and the radicals have easy access to the vessel walls where they recombine. An increase in pressure or the coating of the surface with a reactive material reduces the probability of surface reactions and enhances explosion. If a larger vessel is used, the radicals have further to diffuse to the surface and explosion is more likely to occur.

The pressure at which the second or upper explosion limit occurs is found to be insensitive to these surface parameters and therefore does not depend on surface recombination of the radicals. It is proposed that at high pressures the radicals are removed by recombination in the gas phase. The addition of a foreign or inert gas to the reaction mixture assists the gas-phase recombination and lowers the explosion limit.

7.5.2   *Kinetics of branched chain reactions*

The theory of the kinetics of branched chain reactions is based on the investigations of Hinshelwood in England and Semenov in Russia in the 1930s. Their theory can be illustrated by a simplified treatment using the following general mechanism for a branched chain reaction.

$$I \rightarrow R\cdot \qquad \text{Initiation}$$

$$R\cdot + \ldots \rightarrow P + R\cdot \qquad \text{Propagation}$$

$$R\cdot + \ldots \rightarrow \alpha R\cdot \qquad \text{Branching}$$

$$R\cdot + \ldots \rightarrow ? \qquad \text{Surface termination}$$

$$R\cdot + \ldots \rightarrow ? \qquad \text{Gas-phase termination}$$

where I is an initiator molecule that generates a free radical, $R\cdot$ is any radical and P is the reaction product.

Let v_i be the rate of initiation, and r_p, r_b, r_s and r_g be the rate coefficients for the propagation, branching, surface termination and gas-phase termination processes, respectively. The rate coefficient is the product of a rate constant and a concentration term. For example, one possible propagation process in the hydrogen–oxygen reaction is

$$HO_2\cdot + H_2 \rightarrow H_2O + OH\cdot$$

and the rate is given by $r_p[HO_2\cdot]$, where $r_p = k_p[H_2]$.

Consider the steady-state equation for $R\cdot$

$$\frac{d[R\cdot]}{dt} = v_i + r_b(\alpha - 1)[R\cdot] - r_s[R\cdot] - r_g[R\cdot] = 0 \qquad (7.23)$$

where $\alpha - 1$ is the increase in free radicals in the branching reaction, often equal to two.

From equation 7.23

$$[R\cdot] = \frac{v_i}{r_s + r_g - r_b(\alpha - 1)}$$

The overall rate of reaction if the steady-state approximation holds is therefore

$$v = \frac{d[P]}{dt} = r_p[R\cdot]$$

$$= \frac{r_p v_i}{r_s + r_g - r_b(\alpha - 1)} \qquad (7.24)$$

For steady-state conditions to hold, no branching must occur, that is $\alpha = 1$. When brancing occurs α becomes greater than one and the $r_b(\alpha - 1)$ term increases so that the denominator of equation 7.24 decreases. Therefore as branching increases the rate increases until the denominator becomes equal to zero or the rate becomes infinite. This is the condition for explosion, namely that

$$r_s + r_g = r_b(\alpha - 1) \qquad (7.25)$$

Since steady-state conditions do not apply this is an approximation and in practice the rate becomes very large rather than infinite.

If this theory is applied to the hydrogen–oxygen reaction, the first and second explosion limits can be explained. At low pressures r_s is large so that $r_s + r_g > r_b(\alpha - 1)$. As the pressure is increased r_s decreases until $r_s + r_g = r_b(\alpha - 1)$ when the first explosion limit is observed. At relatively high pressures r_g is high so that $r_s + r_g > r_b(\alpha - 1)$ and the system is stable. When the pressure is reduced r_g decreases until $r_s + r_g = r_b(\alpha - 1)$ again and the second explosion limit is observed.

The occurrence of the third explosion limit is either due to a thermal explosion or to some further branching reaction, which causes a sudden increase in the concentration of free radicals. The nature of the third limit is not well understood.

Problems

1. The following reaction scheme has been postulated for the gas-phase pyrolysis of ethane

$$C_2H_6 = CH_3\cdot + CH_3\cdot$$

$$C_2H_6 + CH_3\cdot = CH_4 + C_2H_5\cdot$$

$$C_2H_5\cdot = C_2H_4 + H\cdot$$

$$C_2H_6 + H\cdot = C_2H_5\cdot + H_2$$

$$C_2H_5\cdot + C_2H_5\cdot = C_4H_{10} \text{ or } C_2H_4 + C_2H_6$$

Derive an expression for the rate of formation of hydrogen in terms of the reactant concentration and the rate constants of the elementary steps in the reaction scheme by application of the steady-state treatment to reactive intermediates.

[University of Edinburgh BSc (2nd Year) 1972]

2. Ethyl chloride undergoes a gas-phase thermal decomposition according to the overall equation

$$C_2H_5Cl \rightarrow C_2H_4 + HCl$$

A possible chain mechanism for this pyrolysis is as follows

$$C_2H_5Cl \rightarrow C_2H_5\cdot + Cl\cdot$$

$$C_2H_5\cdot + C_2H_5Cl \rightarrow C_2H_4Cl\cdot + C_2H_6$$

$$Cl\cdot + C_2H_5Cl \rightarrow C_2H_4Cl\cdot + HCl$$

$$C_2H_4Cl\cdot \rightarrow C_2H_4 + Cl\cdot$$

$$C_2H_4Cl\cdot + Cl\cdot \rightarrow C_2H_4Cl_2$$

Given that the chains are long and that the pressure is high, show that this mechanism leads to an expression for the rate of reaction, which is first order in ethyl chloride, with an overall rate constant k_r, given by $k_r = (k_1 k_2 k_4 / k_5)^{1/2}$.

[University of Hull BSc (Final Exam) 1972]

3. Assuming that gaseous ozone decomposes by the following mechanism

$$O_3 \underset{k_{-1}}{\overset{k_1}{\rightleftharpoons}} O_2 + O:$$

$$O: + O_3 \xrightarrow{k_2} 2O_2$$

derive an expression for the rate of decomposition in the presence of oxygen using a stationary-state method. Show that the mechanism would be consistent with the observation that the reaction is second order in ozone and inhibited by oxygen.

4. Consider the following sequence of reactions for the decomposition of a peroxide ROOR in a solvent SH

$$ROOR \xrightarrow{k_1} 2RO \cdot$$

$$RO \cdot + SH \xrightarrow{k_2} ROH + S \cdot$$

$$S \cdot + ROOR \xrightarrow{k_3} SOR + RO \cdot$$

$$2S \cdot \longrightarrow S_2$$

Show that

$$-\frac{d[ROOR]}{dt} = k_1[ROOR] + k'[ROOR]^{3/2}$$

[University of Lancaster BSc (Part 2) 1971]

5. The solution polymerisation of methyl methacrylate using benzoyl peroxide as initiator and benzene as solvent follows a rate equation of the form

$$\text{rate} = k[M]^x[I]^y$$

where M and I refer to monomer and initiator respectively. From the following set of results determine x and y and outline the kinetic scheme which would explain these results.

Concentration of peroxide /g(100 g monomer)$^{-1}$	Rate of polymerisation /% conversion min^{-1}	Concentration of monomer /mol dm^{-3}	Rate of polymerisation /% conversion min^{-1}
0.04	0.95	0.02	0.40
0.10	1.60	0.03	0.66
0.15	2.00	0.04	0.88
0.20	2.23	0.05	1.06
0.30	2.90	0.06	1.30
0.40	3.21	0.07	1.52
		0.08	1.72

Further reading

References

1. M. Bodenstein and S. C. Lind. *Z. phys. Chem.*, **57** (1906), 168.
2. F. Paneth and W. Hofeditz. *Ber. dt. chem. Ges.*, **B62** (1929), 1335.
3. F. O. Rice and K. F. Herzfeld. *J. Am. chem. Soc.*, **56** (1934), 284.
4. K. J. Laidler and M. T. H. Liu. *Proc. R. Soc.*, **A297** (1967), 365.

Reviews

R. S. Baldwin and R. W. Walker, Branching chain reactions. *Essays in Chem.*, 3, Academic Press (1972).

G. M. Burnett. Rate constants in radical polymerisation reactions. *Q. Rev. chem. Soc.*, **4** (1950), 292.

G. M. Burnett. The study of radical polymerisation in solution. *Progr. Reaction Kinetics*, 3 (1965), 449.

V. N. Kondratiev. Chain reactions. *Comprehensive Chemical Kinetics*, 2 (1969), 81.

V. V. Voevodsky and V. N. Kondratiev. Determination of rate constants for elementary steps in branched chain reactions. *Progr. Reaction Kinetics* 1 (1963) 41.

Books

F. S. Dainton. *Chain Reactions* (2nd edition), Methuen, London (1966).

F. G. R. Gimblett. *Introduction to the Kinetics of Chemical Chain Reactions*, McGraw-Hill, London (1970).

K. J. Laidler. *Reaction Kinetics, Vol. 1, Homogeneous Gas Reactions.* Pergamon, Oxford (1963), chapter 4.

W. A. Pryor. *Free Radicals*, Prentice-Hall, Englewood Cliffs, N.J. (1966).

A. F. Trotman-Dickenson. *Free Radicals*, Methuen, London (1959).

N. N. Semenov. *Chemical Kinetics and Chain Reactions,* Clarendon Press, Oxford (1935).

8 REACTIONS IN SOLUTION

8.1 Comparison between reactions in gas phase and in solution

Reactions in the gas phase can largely be considered in terms of the kinetic theory of gases, and involve isolated collisions between individual molecules. Liquids are not as simple as either solids or gases and consequently the theory of reactions in solution is rather complex. Nevertheless since most chemical reactions of importance occur in solution, a knowledge of such reactions, even if based on empirical factors, is highly desirable.

The main difference between gas- and liquid-phase reactions is that in the latter reactant molecules collide continuously with solvent molecules. In systems in which the solvent has little or no effect on the rate, the rate and the mechanism do not differ much from the corresponding gas-phase reaction. However in many reactions in solution the presence of a solvent results in ionisation, so that reactions between ions can be studied. Since the rate depends on the electrical environment of the ions, it is influenced by the dielectric constant of the solvent.

Another difference observed when reactions occur in solution is that there is a greater number of collisions per unit time. Energy transfer is rapid, and thermal and vibrational equilibrium are attained very rapidly. Also, Rabinovitch[1] showed from simulation experiments that collisions occur in sets when the molecules are closely packed together. With such molecules, once an initial collision has occurred the surrounding molecules form a 'cage', within which, a large number of subsequent collisions take place, before the molecules separate. This 'cage effect' is important in processes occurring with a low activation energy such as a combination between two free radicals. A zero activation energy implies that chemical reaction occurs at every collision. This phenomenon results in reaction at the first collision followed by a number of collisions within the cage, which do not contribute to the rate. If a reactant molecule is photochemically decomposed in solution (this is a temperature-independent process and therefore has zero activation energy) the resultant free radicals often recombine within the cage of the surrounding solvent molecules before separation takes place.

In reactions with low activation energy (usually less than 20 kJ mol^{-1}), the rate-determining step could well be the rate of diffusion of the reactant molecules to each other resulting in the collision, or the rate of diffusion of products away from each other after collision. Such reactions are said to be *diffusion-controlled* and their rates will depend on the viscosity of the solvent. Free-radical recombination reactions in solution are therefore always diffusion-controlled.

Other reactions in solution are different from gas-phase processes because the solvent is involved chemically in the mechanism. In some cases it may act as a catalyst, while in other reactions it is consumed during the reaction.

8.2   Transition state theory for liquid reactions

For bimolecular gas reactions the rate constant for the reaction

$$A + B \rightarrow X^{\ddagger} \rightarrow \text{products}$$

is (see page 65)

$$k_r = \nu \exp\left(-\Delta G^{\ddagger}/RT\right) \tag{8.1}$$

that is

$$k_r = \frac{kT}{h} K^{\ddagger} \tag{8.2}$$

giving

$$k_r = \frac{kT}{h} \exp\left(-\frac{\Delta G^{\ddagger}}{RT}\right) \tag{8.3}$$

and

$$k_r = \frac{kT}{h} \exp\left(\frac{\Delta S^{\ddagger}}{R}\right) \exp\left(-\frac{\Delta H^{\ddagger}}{RT}\right) \tag{8.4}$$

where $\Delta G^{\ddagger}$, $\Delta S^{\ddagger}$ and $\Delta H^{\ddagger}$ are the free energy of activation, the entropy of activation and the enthalpy of activation, respectively.

Although these thermodynamic properties can be obtained from the literature, the effect of the solvent makes their values rather uncertain. While accurate data are not available it has been found, as in gas-phase reactions, that the entropy of activation gives a useful indication of the structure of the transition state. A positive entropy of activation indicates that the transition state is less ordered than the individual reactant molecules, while a negative entropy of activation corresponds to an increase in order when the reactant molecules combine to form the transition state.

Entropies of activation can be experimentally determined, and a number of attempts have been made to estimate these values from theoretical considerations. In recent years reactions in solution have been studied at high pressures and their volumes of activation calculated as described in section 8.4.

It is found that the change in volume when the activated state is formed can be correlated with the entropy of activation. Therefore reactions with negative $\Delta S^{\ddagger}$ and $\Delta V^{\ddagger}$ values are always slower than normal, while reactions with positive $\Delta S^{\ddagger}$ and $\Delta V^{\ddagger}$ values are faster than normal. Since changes in volume are very sensitive to changes in the electrical environment of the reactants, reactions that proceed via a similar mechanism will have similar $\Delta V^{\ddagger}$ values. Considerable progress has been made using $\Delta V^{\ddagger}$ values as a guide to the reaction mechanism.

8.3 Reactions involving ions

Reactions between ions are often too rapid to be measured by conventional methods, and can only be studied by methods described in chapter 11. For instance, the reaction

$$H^+ + OH^- \rightarrow H_2O$$

in the neutralisation of strong acids and bases is one of the fastest reactions with a rate constant of $1.4 \times 10^{11} \, dm^3 \, mol^{-1} \, s^{-1}$. However, many reactions between ions—particularly processes involving the breakage and formation of a covalent bond—proceed at measurable rates. For example, the reaction

$$CH_3Br + Cl^- \rightarrow CH_3Cl + Br^-$$

in acetone at 298 K has a rate constant of $5.9 \times 10^{-3} \, dm^3 \, mol^{-1} \, s^{-1}$.

Many experimental and other parameters have been shown to influence the rates of ionic reactions, but for simplicity discussion will be restricted to the effect of the following on reaction rates:

(1) the nature of the solvent;
(2) the nature of the ions;
(3) the ionic strength of the solution.

8.3.1 The nature of the solvent

Equation 8.3 shows that the rate constant of a reaction depends on the free energy of activation $\Delta G^{\ddagger}$. In an ionic reaction the electrostatic interaction between the ions makes an important contribution to this free energy of activation. Since this is a measure of the change in free energy in going from the reactant state to the activated state, the electrostatic contribution to the free energy of activation $\Delta G_{e.s.}^{\ddagger}$ depends on the structure formulated for the activated state.

Two approaches have been suggested as follows. In both cases it is assumed that the ions act as conducting spheres with charges z_A and z_B, respectively, in a solvent which is regarded as a continuous dielectric with a fixed dielectric constant ϵ. If the activated complex forms as a double sphere as shown in figure 8.1, the rate constant k_r is given by the relationship

$$\ln k_r = \ln k_0 - \frac{z_A z_B e^2}{\epsilon\, d_{AB} kT}$$

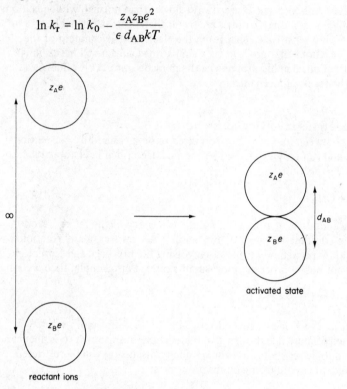

Figure 8.1   *Model for double-sphere activated complex*

where k_0 is the rate constant in a medium of infinite dielectric constant, e is the charge of an electron and d_{AB} is the 'internuclear distance' in the activated complex. Therefore

$$\log_{10} k_r = \log_{10} k_0 - \frac{z_A z_B e^2}{2.303\epsilon\, d_{AB} kT} \tag{8.5}$$

Alternatively if a single-sphere model for the activated complex is proposed as illustrated by figure 8.2 a rather more complex expression results

$$\ln k_r = \ln k_0 - \frac{e^2}{2\epsilon kT}\left[\frac{(z_A + z_B)^2}{r_\ddagger} - \frac{z_A^2}{r_A} - \frac{z_B^2}{r_B}\right]$$

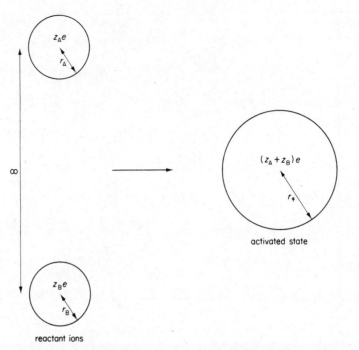

Figure 8.2 *Model of a single-sphere activated complex*

that is

$$\log_{10} k_r = \log_{10} k_0 - \frac{e^2}{4.606 \epsilon kT} \left[\frac{(z_A + z_B)^2}{r_\ddagger} - \frac{z_A^2}{r_A} - \frac{z_B^2}{r_B} \right] \tag{8.6}$$

This equation reduces to equation 8.5 when $r_A = r_B = r_\ddagger$.

It is important to note that both approaches predict that a plot of $\log_{10} k_r$ against $1/\epsilon$ will be linear. If the double-sphere model is applicable, the slope is equal to $z_A z_B e^2 / 2.303 d_{AB} kT$. A typical plot for the reaction between azodicarbonate ions and hydrogen ions in water–dioxane solutions is shown in figure 8.3. This method has proved to be a useful method for the estimation of d_{AB} values. A linear plot is obtained except at low ϵ values, and for the above reaction d_{AB} is 340 pm.

8.3.2 The nature of the ions

The nature of the activated complex as formulated by a double-sphere model is expected to depend on the charge of the reacting ions. Consider a reaction between two negatively charged ions. The activated complex is

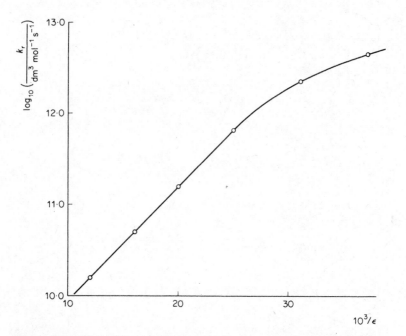

Figure 8.3 *Plot of* $\log_{10} k_r$ *against* $1/\epsilon$ *for the reaction between azodicarbonate ions and hydrogen ions at* $25^\circ C$

doubly charged and the solvent molecules near this complex are subject to strong electrostatic forces. As a result the solvent molecules are less free to move in the region of the complex and a decrease in entropy occurs. This phenomenon is known as *electrostriction* and results in a negative entropy of activation. The same is true for a reaction between two positively charged ions.

In a reaction between two ions of opposite charge, the charge associated with the activated complex decreases. This results in a decrease in electrostriction and a positive entropy of activation.

Since the frequency factor in the Arrhenius equation is proportional to $\exp(\Delta S^\ddagger/R)$ (equation 5.18) the above theory postulates that for reaction between ions of like charge, the frequency factor is less than normal. Conversely, for reactions between ions of opposite charge, the frequency factor is greater than normal. Inspection of table 8.1 shows that the experimental A factors for a number of ionic reactions are consistent with this theory.

It is assumed in the absence of electrostatic effects that the entropy of activation is zero and the frequency factor is 'normal', that is of the order of 10^{12} dm^3 mol^{-1}s^{-1}.

TABLE 8.1 SOME A FACTORS AND ENTROPIES OF
ACTIVATION FOR SOME REACTIONS BETWEEN
IONS

Reactants	A /dm^3 mol^{-1} s^{-1}	$\Delta S^{\ddagger}$ /J K^{-1} mol^{-1}
	Experimental	
$Co(NH_3)_5 Br^{2+} + Hg^{2+}$	1×10^8	-100
$CH_2 BrCOO^- + S_2O_3{}^{2-}$	1×10^9	-71
$CH_2 ClCOO^- + OH^-$	6×10^{10}	-50
$CH_2 BrCOOCH_3 + S_2O_3{}^{2-}$	1×10^{14}	$+25$
$Co(NH_3)_5 Br^{2+} + OH^-$	5×10^{17}	$+92$

8.3.3 Ionic strength of the solution

Brønsted,[2] Bjerrum[3] and others[4] showed that the rate of an ionic reaction
depends on the ionic strength of the solution. Since the ionic strength of a
solution can be changed by the addition of an inert salt, this is known as
the *primary salt effect*. The theoretical basis for the influence of the ionic
strength on the rate constant is derived as follows.

Consider an ionic reaction

$$A + B \rightarrow X^{\ddagger} \rightarrow products$$

The equilibrium constant for this reaction is defined in terms of relative
activities. The relative activity a of a solution is given by

$$a = \gamma c$$

where c is the concentration and γ is the activity coefficient. The equilibrium
constant K is given by

$$K = \frac{a_{X^{\ddagger}}}{a_A a_B} = \frac{[X^{\ddagger}]}{[A][B]} \frac{\gamma_{X^{\ddagger}}}{\gamma_A \gamma_B} \tag{8.7}$$

Therefore

$$[X^{\ddagger}] = K[A][B] \frac{\gamma_A \gamma_B}{\gamma_{X^{\ddagger}}} \tag{8.8}$$

It is assumed that the rate of the above reaction depends only on the concentration of the activated complex so that the rate of reaction v is given by

$$v = -\frac{d[A]}{dt} = -\frac{d[B]}{dt} = k'[X^{\ddagger}] \tag{8.9}$$

Substitution of equation 8.8 in equation 8.9 gives

$$v = k'K[\text{A}][\text{B}] \frac{\gamma_\text{A}\gamma_\text{B}}{\gamma_\text{X}\ddagger} \tag{8.10}$$

But for this reaction, the rate v and the rate constant k_r are related by

$$v = k_\text{r}[\text{A}][\text{B}] \tag{8.11}$$

Combining equations 8.10 and 8.11 gives

$$k_\text{r} = k'K \frac{\gamma_\text{A}\gamma_\text{B}}{\gamma_\text{X}\ddagger} \tag{8.12}$$

Let k_0 be the rate constant at infinite dilution (zero ionic strength) when the activity coefficients are equal to unity. Therefore under these conditions

$$k_0 = k'K$$

giving in general

$$k_\text{r} = k_0 \frac{\gamma_\text{A}\gamma_\text{B}}{\gamma_\text{X}\ddagger} \tag{8.13}$$

Taking logarithms

$$\log_{10} k_\text{r} = \log_{10} k_0 + \log_{10}\left(\frac{\gamma_\text{A}\gamma_\text{B}}{\gamma_\text{X}\ddagger}\right) \tag{8.14}$$

From the Debye–Hückel limiting law, the activity coefficient of an ion i of charge z_i is related to the ionic strength I by

$$\log_{10} \gamma_\text{i} = -Az_\text{i}^2 \sqrt{I}$$

where A is the Debye–Hückel constant and the ionic strength $I = \Sigma \frac{1}{2}c_\text{i}z_\text{i}^2$. Therefore

$$\log_{10} \frac{\gamma_\text{A}\gamma_\text{B}}{\gamma_\text{X}\ddagger} = -A\sqrt{I}[z_\text{A}^2 + z_\text{B}^2 - (z_\text{A} + z_\text{B})^2]$$

since the charge on the complex is the sum of the two charges on the reacting ions; that is

$$\log_{10} \frac{\gamma_\text{A}\gamma_\text{B}}{\gamma_\text{X}\ddagger} = +2Az_\text{A}z_\text{B}\sqrt{I} \tag{8.15}$$

Substituting equation 8.15 into equation 8.14 gives

$$\log_{10} k_\text{r} = \log_{10} k_0 + 2Az_\text{A}z_\text{B}\sqrt{I} \tag{8.16}$$

This is known as the *Brønsted–Bjerrum relationship* and predicts that a plot of $\log_{10} k_r$ against $\sqrt{I}$ is linear with a slope equal to $2Az_A z_B$ and the intercept is equal to $\log_{10} k_0$. For aqueous solutions at $25°C$, the Debye–Hückel constant $A = 0.51$ $dm^{3/2}$ $mol^{-1/2}$.

Example 8.1
The second-order reaction between bromoacetate ions and thiosulphate ions in which the potassium salts were used

$$BrCH_2COO^- + S_2O_3^{2-} \to S_2O_3CH_2COO^{2-} + Br^-$$

was carried out with equal concentrations of both reactants. The rate constants were as follows

Rate constant k_r / dm^3 mol^{-1} min^{-1}	0.298	0.309	0.324	0.343	0.366
Initial concentration/mol dm^{-3} $\times 10^{-3}$	0.5	0.7	1.0	1.4	2.0

Calculate the rate constant when the activity coefficients are unity, assuming the Debye–Hückel constant = 0.51 $dm^{3/2}$ $mol^{-1/2}$.

The Brønsted–Bjerrum relationship (equation 8.16) predicts that a plot of $\log_{10} k_r$ against $\sqrt{I}$ will be linear. The ionic strength of the reaction mixture is given by $I = \frac{1}{2} \sum c_i z_i^2$; that is

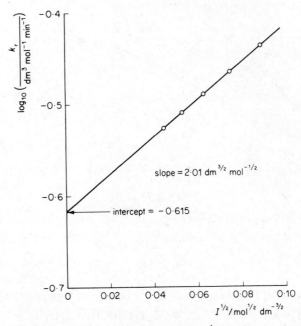

Figure 8.4 *Plot of $\log_{10} k_r$ against $\sqrt{I}$ for the reaction between bromoacetate and thiosulphate ions*

$$I = \tfrac{1}{2}[c_{K^+} \times 1^2 + c_{BrAc^-} \times 1^2 + 2c_{K^+} \times 1^2 + c_{S_2O_3^{2-}} \times 2^2]$$

where c is the concentration and z the charge of each ion i in the mixture.

$\log_{10}k_r/dm^3\,mol^{-1}\,min^{-1}$	-0.526	-0.510	-0.490	-0.465	-0.437
$I/mol\,dm^{-3}$	0.0020	0.0028	0.0040	0.0056	0.0080
$\sqrt{I}/mol\,dm^{-3}$	0.0447	0.0529	0.0633	0.0748	0.0894

A plot of $\log_{10} k_r$ against $\sqrt{I}$ is given in figure 8.4. The graph is linear with slope = 2.01 $dm^{3/2}\,mol^{-1/2}$, which is in good agreement with the predicted value for $2Az_Az_B$ of $2 \times 2 \times 0.51\,dm^{3/2}\,mol^{-1/2} = 2.04\,dm^{3/2}\,mol^{-1/2}$, and intercept = $\log_{10}k_0 = -0.615 = \bar{1}.385$. This gives

$$k_0 = 0.243\,dm^3\,mol^{-1}\,min^{-1}$$

Rearrangement of equation 8.16 gives

$$\log_{10}\left(\frac{k_r}{k_0}\right) = 2Az_Az_B\sqrt{I} \tag{8.17}$$

Therefore a plot of $\log_{10} k_r/k_0$ against $\sqrt{I}$ is linear. Figure 8.5 shows this plot for the ionic reactions listed in table 8.1. It is seen that for reactions

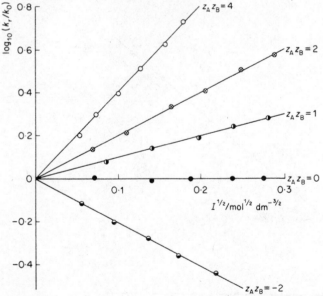

Figure 8.5 *Variation of $\log_{10} k/k_0$ with $\sqrt{I}$ for a number of ionic reactions*

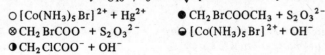

O $[Co(NH_3)_5\,Br]^{2+} + Hg^{2+}$ ● $CH_2\,BrCOOCH_3 + S_2O_3^{2-}$

⊗ $CH_2\,BrCOO^- + S_2O_3^{2-}$ ◒ $[Co(NH_3)_5\,Br]^{2+} + OH^-$

◑ $CH_2\,ClCOO^- + OH^-$

between ions of like charge the slope is positive. Such reactions show a positive salt effect; that is, the reaction rate increases with increasing ionic strength. For reactions between ions of opposite charge the slope is negative. This corresponds to a negative salt effect and the reaction rate decreases with increasing ionic strength. A reaction between an ion and a neutral molecule such as the acid or alkaline hydrolysis of an ester does not exhibit a primary salt effect.

8.4 Effect of pressure on reaction rates

The field of high-pressure chemistry is still in its early years and apparatus capable of measuring reaction rates over a wide range of pressure is rather expensive. While the average student is likely to have investigated the effect of temperature on a reaction rate in his laboratory programme, he is unlikely to have access to equipment that measures the effect of pressure on reaction rates. Nevertheless the information gained from such experiments has proved invaluable to the understanding of many reaction mechanisms. In particular it provides a method of measuring volumes of activation.

From the van't Hoff equation

$$\Delta G = -RT \ln K$$

But volume and free energy are related by

$$V = \left(\frac{\partial G}{\partial p} \right)_T$$

or

$$\Delta V = \left(\frac{\partial (\Delta G)}{\partial p} \right)_T = -RT \left(\frac{d \ln K}{dp} \right)_T$$

giving

$$\left(\frac{d \ln K}{dp} \right)_T = -\frac{\Delta V}{RT}$$

The volume of activation $\Delta V^{\ddagger}$ is defined as the change in volume in going from the reactant state to the activated state. Therefore

$$\left(\frac{d \ln K^{\ddagger}}{dp} \right)_T = -\frac{\Delta V^{\ddagger}}{RT} \tag{8.18}$$

Since $k_r = (kT/h)K^{\ddagger}$, the variation of rate constant with pressure is given by

$$\left(\frac{d \ln k_r}{dp} \right)_T = -\frac{\Delta V^{\ddagger}}{RT} \tag{8.19}$$

This relationship predicts that if the rate constant increases with increasing pressure, the volume of the activated state is less than that of the reactants; that is, $\Delta V^{\ddagger}$ is negative. Conversely, a rate constant that decreases with increasing pressure corresponds to a positive $\Delta V^{\ddagger}$.

Volumes of activation can be determined from measurements of the rate constant carried out over a wide range of pressure at constant temperature. Integration of equation 8.19 gives

$$\ln k_{\mathrm{r}} = -\frac{\Delta V^{\ddagger}}{RT} p + \text{constant}$$

Let k_0 be the rate constant for the reaction at zero pressure, or within experimental error, atmospheric pressure. Therefore, constant $= \ln k_0$, and

$$\ln k_{\mathrm{r}} = \ln k_0 - \frac{\Delta V^{\ddagger}}{RT} p$$

that is

$$\log_{10} k_{\mathrm{r}} = \log_{10} k_0 - \frac{\Delta V^{\ddagger}}{2.303RT} p \tag{8.20}$$

A plot of $\log_{10} k_{\mathrm{r}}$ against p is therefore linear and the slope is given by $\Delta V^{\ddagger}/2.303RT$, from which $\Delta V^{\ddagger}$ is determined.

Example 8.2
The following data were obtained by Laidler and Chen (*Trans. Faraday Soc.* **54** (1958), 1026) for the alkaline hydrolysis of methyl acetate at $25°C$

Pressure/MN m^{-2}	0.1	27.6	55.2	82.7
Rate constant k_{r}/dm^3 mol^{-1} s^{-1}	0.146	0.163	0.181	0.203

Calculate the volume of activation.

Equation 8.20 predicts that a plot of $\log_{10} k_{\mathrm{r}}$ against p will be linear.

$\log_{10} k_{\mathrm{r}}$/dm^3 mol^{-1} s^{-1}	$\bar{1}.164$	$\bar{1}.212$	$\bar{1}.258$	$\bar{1}.308$
p/MN m^{-2}	0.1	27.6	55.2	82.7

The graph is drawn in figure 8.6. From the graph

$$\text{slope} = -\frac{\Delta V^{\ddagger}}{2.303RT} = 1.74 \times 10^{-3} \text{ m}^2 \text{ (MN)}^{-1}$$

that is

$$\Delta V^{\ddagger} = -1.74 \times 10^{-3} \times 10^{-6} \times 2.303 \times 8.314 \times 298 \text{ m}^3 \text{ mol}^{-1}$$
$$= -9.93 \times 10^{-6} \text{ m}^3 \text{ mol}^{-1} = -9.93 \text{ cm}^3 \text{ mol}^{-1}$$

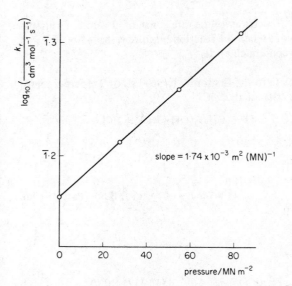

Figure 8.6 *Variation of rate constant with pressure for the alkaline hydrolysis of methyl acetate*

Problems

1. The reaction between persulphate ions and iodide ions is as follows

$$S_2O_8^{2-} + 2I^- \rightarrow 2SO_4^{2-} + I_2$$

Using an initial concentration of 1.5×10^{-4} mol dm^{-3} potassium persulphate, King and Jacobs (*J. Am. chem. Soc.*, **53** (1931), 53) obtained the following values of the rate constant k_r at the concentrations of potassium iodide given

10^3[KI]/mol dm^{-3}	1.6	2.0	3.2	4.0	6.0	8.0	10.00
$10\,k_r$/dm^3 mol^{-1} min^{-1}	1.03	1.05	1.12	1.16	1.18	1.26	1.32

Estimate $z_A z_B$ for this reaction and calculate the rate constant when the activity coefficients are unity.

2. Brønsted and Livingstone (*J. Am. chem. Soc.*, **49** (1927), 435) found a rate constant at 15°C of 1.52 dm^3 mol^{-1} s^{-1} for the reaction

$$[CoBr(NH_3)_5]^{2+} + OH^- \rightarrow [Co(NH_3)_5OH]^{2+} + Br^-$$

measured during an experiment when the initial concentration of $[CoBr(NH_3)_5]^{2+}$ ion (present in the form of the bromide) was 5×10^{-4} mol dm^{-3} and the OH$^-$ ion concentration (present as sodium hydroxide) was

$7.05 \times 10^{-4} \, \text{mol dm}^{-3}$. Calculate (a) the rate constant at zero ionic strength, and (b) the rate constant when the reaction mixture is made $5 \times 10^{-3} \, \text{mol dm}^{-3}$ with respect to sodium chloride.

3. The data below relate to the first-order hydrolysis of 2-chloro-2-methyl-but-3-yne in aqueous ethanol at $25°C$.

$$(CH_3)_2CClC\equiv CH + H_2O \rightarrow (CH_3)_2(OH)C\equiv CH + HCl$$

Rate constant $\times 10^6 / \text{s}^{-1}$	0.23	0.38	0.74	1.27	2.04
Pressure/MN m^{-2}	0	107	213	317	421

Calculate the volume of activation.

[University of Salford, B.Sc. (Part 2) 1968]

Further reading

References
1. E. Rabinovitch. *Trans. Faraday Soc.*, **33** (1937), 1225.
2. J. N. Brønsted. *Z. phys. Chem.*, **102** (1922), 169.
3. N. Bjerrum. *Z. phys. Chem.*, **108** (1924), 82.
4. J. A. Christiansen. *Z. phys. Chem.*, **113** (1924), 35; G. Scatchard. *Chem. Rev.*, **10** (1932), 229.

Reviews
C. W. Davies. Salt effects in solution kinetics. *Progr. Reaction Kinetics*, **1** (1963), 161.
B. Perlmutter-Hayman. The primary kinetic salt effect in aqueous solution. *Progr. Reaction Kinetics*, **6** (1971), 239.
C. H. Rochester. Salt and medium effects on reaction rates in concentrated solutions of acids and bases. *Progr. Reaction Kinetics*, **6** (1971), 143.
G. Kohnstam. The kinetic effects of pressure. *Progr. Reaction Kinetics*, **5** (1970), 335.

Books
E. A. Moelwyn-Hughes. *Chemical Statics and Kinetics of Solution*, Academic Press, London (1971).
E. A. Moelwyn-Hughes. *Kinetics of Reactions in Solution*, Clarendon Press, Oxford (1947).
S. D. Hamann. *Physico-Chemical Effects of Pressure*, Butterworths, London (1959), ch. 9.

9 CATALYSED REACTIONS

The addition of a catalyst to a chemical reaction increases the rate of reaction. Catalysts are very important in chemical industry, where their use can increase the efficiency of a chemical process and decrease the overall cost to the manufacturer. It is therefore not surprising that much time and considerable resources have been devoted to the discovery of new and better catalysts. Despite this the mechanisms of many catalysed reactions are not well understood, and many of the catalysts used in chemical industry were discovered by trial and error rather than by fundamental research.

It is important to stress that the addition of a catalyst does not directly affect the thermodynamics of the reaction. The role of the catalyst is to accelerate the forward and reverse reactions to the same extent and thereby reduce the time for equilibrium to be reached. Therefore, in the presence of a catalyst, it is often possible to use different experimental conditions such that the reaction proceeds faster even though the thermodynamics are bad. For example, in the Haber process

$$N_2 + 3H_2 \rightleftharpoons 2NH_3$$

the presence of the catalyst does not change the equilibrium constant, but speeds up the rate of attainment of equilibrium. The formation of ammonia is an exothermic process and therefore favoured by low temperatures. In the presence of the catalyst a higher temperature can be used, and it is found that an economical yield of ammonia is obtained at $450°C$.

The role of the catalyst is seen from the potential energy–reaction path diagram shown in figure 9.1. If $E_1^{\ddagger}$ and $E_{-1}^{\ddagger}$ are the activation energies of the uncatalysed forward and reverse reactions respectively, the catalyst reduces the height of the activation energy barrier by a value E. The corresponding activation energies for the catalysed reactions are $E_1^{\ddagger}(\text{cat})$ and $E_{-1}^{\ddagger}(\text{cat})$ respectively, so that

$$E_1^{\ddagger}(\text{cat}) = E_1^{\ddagger} - E$$
$$E_{-1}^{\ddagger}(\text{cat}) = E_{-1}^{\ddagger} - E$$

that is, the activation energy for the forward and reverse reactions are reduced by the same extent.

The addition of a substance called a catalytic poison or an inhibitor

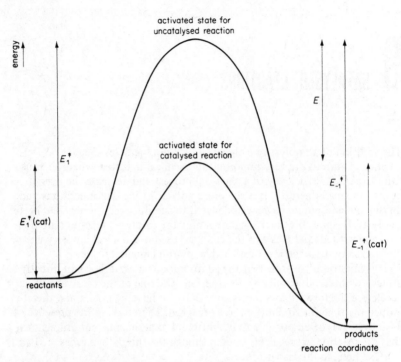

Figure 9.1 *Potential-energy diagram illustrating lower activation-energy barrier for a catalysed reaction*

decreases the rate of a chemical reaction. The term negative catalyst is often used, but this is misleading since the additive is often consumed in the reaction. The inhibitor can slow down a reaction by undergoing a competing process with an intermediate or the catalyst itself.

9.1 Homogeneous catalysis
In this type of catalysed reaction, the reactant(s) and catalyst are present in the same phase and the walls of the vessel do not affect the rate of the reaction. Therefore, if glass beads or a chemically inert solid are added to increase the surface-to-volume ratio and no change in rate occurs, the reaction is said to be homogeneous. Some examples of reactions in the gas phase and the important topic of acid–base catalysis will be considered.

9.1.1 Gas–phase reactions
Nitrogen dioxide has been shown to be an effective catalyst for a number of gas-phase reactions. For example the oxidation of carbon monoxide

$$CO + \tfrac{1}{2}O_2 \rightarrow CO_2$$

in the presence of nitrogen dioxide proceeds as follows

$$CO + NO_2 \rightarrow CO_2 + NO$$

$$NO + \tfrac{1}{2}O_2 \rightarrow NO_2$$

Iodine vapour is also known to catalyse a number of organic pyrolyses. The iodine-catalysed decomposition of acetaldehyde is a chain reaction, which is thought to proceed as follows

$$I_2 \underset{k_{-1}}{\overset{k_1}{\rightleftharpoons}} 2I\cdot \qquad (1)$$

$$I\cdot + CH_3CHO \xrightarrow{k_2} CH_3CO\cdot + HI \qquad (2)$$

$$CH_3CO\cdot \xrightarrow{k_3} CH_3\cdot + CO \qquad (3)$$

$$CH_3\cdot + I_2 \xrightarrow{k_4} CH_3I + I\cdot \qquad (4)$$

$$CH_3\cdot + HI \xrightarrow{k_5} CH_4 + I\cdot \qquad (5)$$

$$CH_3I + HI \xrightarrow{k_6} CH_4 + I_2 \qquad (6)$$

Application of the steady-state approximation to the intermediates gives the rate equation

$$-\frac{d[CH_3CHO]}{dt} = k[I_2]^{1/2}[CH_3CHO]$$

where $k = (k_1/2k_{-1})^{1/2}k_2$.

This mechanism can be compared with the non-catalysed mechanism given on page 86. It is seen that the initiation process gives iodine atoms, which propagate the chain by reaction (2). The net result is that the major products, CH_4 and CO, are the same as in the uncatalysed decomposition. The catalyst, iodine, is regenerated in reaction (6).

By similar considerations as those used on page 89 the activation energy for this reaction is given by

$$E^{\ddagger} = \tfrac{1}{2}(E_1^{\ddagger} - E_6^{\ddagger}) + E_2^{\ddagger}$$

Since the I—I bond is much weaker than the C—C bond in acetaldehyde, initiation is much easier and $E_1^{\ddagger}$ is 204 kJ mol^{-1} (compared with 332 kJ mol^{-1} for the non-catalysed reaction). This value of $E_1^{\ddagger}$ gives an overall activation energy of 134 kJ mol^{-1} for the catalysed decomposition compared with 198 kJ mol^{-1} for the non-catalysed decomposition. The activation-energy barrier as shown in figure 9.1 is therefore lowered by 64 kJ mol^{-1}.

9.2   Acid–base catalysis

Many homogeneous reactions in solution are catalysed by acids and bases. The hydrolysis of an ester is a well-known example of a reaction that is catalysed by either acids or bases. The mutarotation of glucose is an example of a reaction that is catalysed by acids, bases and solvents.

Consider a substrate S which undergoes an elementary reaction with an acid or base or both. The Lowry–Brønsted definition of acids and bases defines an acid as a substance that donates a proton

$$HA + H_2O \rightarrow H_3O^+ + A^-$$

and a base as a substance that accepts a proton

$$A^- + H_2O \rightarrow HA + OH^-$$

The rate of a catalysed reaction is given by

$$v = k_{cat}[S]$$

where k_{cat} is a catalytic rate coefficient. Therefore

$$k_{cat} = k_0 + k_{H^+}[H_3O^+] + k_{OH^-}[OH^-] + k_{HA}[HA] + k_{A^-}[A^-] \quad (9.1)$$

where k_0 is the rate coefficient for the uncatalysed reaction, and k_{H^+}, k_{OH^-} k_{HA} and k_{A^-} are the catalytic rate constants for the species indicated.

Two types of acid–base catalysis have been observed: specific acid–base catalysis and general acid–base catalysis.

9.2.1   Specific acid–base catalysis

There are certain reactions in which the rate is proportional only to the concentration of the H_3O^+ and OH^- ions present. Such reactions are examples of *specific acid–base catalysis*. In this situation either or both k_{H^+} and k_{OH^-} are large compared to k_{HA} and k_{A^-} and equation 9.1 reduces to

$$k_{cat} = k_0 + k_{H^+}[H_3O^+] + k_{OH^-}[OH^-] \quad (9.2)$$

If the reaction is catalysed only by acids as in the inversion of sugar

$$k_{cat} = k_0 + k_{H^+}[H_3O^+] \quad (9.3)$$

Similarly, for a reaction catalysed only by bases

$$k_{cat} = k_0 + k_{OH^-}[OH^-] \quad (9.4)$$

The catalytic rate coefficient is determined by measuring the rate in a solution of constant ionic strength over a wide range of pH with appropriate buffers.

Consider a reaction with a high acid concentration when equation 9.3 becomes

$$k_{\text{cat}} = k_{\text{H}^+}[\text{H}_3\text{O}^+] \qquad (9.5)$$

Taking logarithms

$$\log_{10} k_{\text{cat}} = \log_{10} k_{\text{H}^+} + \log_{10}[\text{H}_3\text{O}^+]$$

or

$$\log_{10} k_{\text{cat}} = \log_{10} k_{\text{H}^+} - \text{pH} \qquad (9.6)$$

In the acid pH range a plot of $\log_{10} k_{\text{cat}}$ against pH is linear with a slope equal to -1.

In a strongly basic solution equation 9.4 becomes

$$k_{\text{cat}} = k_{\text{OH}^-}[\text{OH}^-] = k_{\text{OH}^-} \frac{K_{\text{w}}}{[\text{H}_3\text{O}^+]} \qquad (9.7)$$

where K_{w} is the ionic product of water.

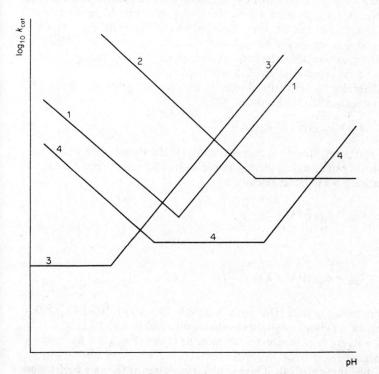

Figure 9.2 *Effect of pH on rate of some acid-base catalysed reactions*

Taking logarithms

$$\log_{10} k_{cat} = \log_{10} k_{OH^-} K_w + pH \tag{9.8}$$

A plot of $\log_{10} k_{cat}$ against pH is therefore linear with a slope equal to +1.

In the region of intermediate pH $\log_{10} k_{cat}$ becomes independent of $[H_3O^+]$ and $[OH^-]$. In such cases k_{cat} depends only on k_0 and is independent of pH.

The effects of pH on $\log_{10} k_{cat}$ for some examples of reactions subject to specific acid-base catalysis are given in figure 9.2. The hydrolysis of esters is represented by curve 1 and is seen to be subject to both acid and base catalysis. The inversion of sugar is catalysed by acids only as shown by curve 2; the aldol condensation of acetaldehyde is catalysed by bases only as shown by curve 3; the mutarotation of glucose shown by curve 4 has a range of pH over which the rate is unaffected by acid or base catalysts.

9.2.2 General acid–base catalysis

Reactions that are catalysed by all Lowry-Brønsted acids and bases in the solution are said to exhibit *general acid–base catalysis*. If the solution is buffered so that the rate is not affected by H_3O^+ and OH^- ions—that is, k_{H^+} and k_{OH^-} are negligible and constant ionic strength is maintained—the rate of many reactions depends on the concentration of undissociated acid HA and the concentration of the conjugate base A^-.

Assuming the uncatalysed reaction makes a negligible contribution to the rate, equation 9.1 becomes

$$k_{cat} = k_{HA}[HA] + k_{A^-}[A^-] \tag{9.9}$$

The reaction is studied at two different pH values where $[HA]/[A^-]$ is a known constant, say x_1 and x_2, respectively. Under these conditions, equation 9.9 can be represented by

$$k_{cat} = k_{HA}[HA] + k_{A^-} \frac{[HA]}{x_1}$$

and

$$k_{cat} = k_{HA}[HA] + k_{A^-} \frac{[HA]}{x_2}$$

A plot of k_{cat} against [HA] for solutions buffered with $[HA]/[A^-]$ ratios of x_1 and x_2 gives straight lines with slopes equal to $k_{HA} + k_{A^-}/x_1$ and $k_{HA} + k_{A^-}/x_2$ respectively, from which the values of k_{HA} and k_{A^-} are determined.

Using these methods it is possible to evaluate the five rate coefficients given in equation 9.1 and therefore derive k_{cat} for a catalysed reaction at a given temperature and pH. For example, Bell and Jones have shown that for the iodination of acetone

$$k_0 = 5 \times 10^{-10}\,s^{-1}$$

$$k_{H^+} = 1.6 \times 10^{-3}\,dm^3\,mol^{-1}\,s^{-1}$$

$$k_{OH^-} = 15\,dm^3\,mol^{-1}\,s^{-1}$$

$$k_{HA} = 5 \times 10^{-6}\,dm^3\,mol^{-1}\,s^{-1}$$

and

$$k_{A^-} = 15 \times 10^{-6}\,dm^3\,mol^{-1}\,s^{-1}$$

9.2.3 Brønsted catalysis law

The ionisation constant of an acid or base is a measure of the strength of the acid or base, and is therefore expected to be a measure of its efficiency as a catalyst. The relationship between the catalytic rate coefficient and the ionisation constant of an acid or base is shown by the Brønsted catalysis law. For an acid

$$k_{cat} = G_a K_a{}^\alpha \tag{9.10}$$

where K_a is the ionisation constant and α and G_a are constants usually having values between zero and unity. The constants are characteristic for a reaction in a given solvent at a given temperature. Similarly for a base

$$k_{cat} = G_b K_b{}^\beta \tag{9.11}$$

where K_b is the ionisation constant of the base and β and G_b are constants.

If there are p groups on the catalyst that can donate protons and q groups that can accept protons, the relationship for general acid catalysis becomes

$$\frac{k_{cat}}{p} = G_a \left(\frac{q}{p} K_a \right)^\alpha \tag{9.12}$$

and for general base catalysis

$$\frac{k_{cat}}{q} = G_b \left(\frac{p}{q} K_b \right)^\beta \tag{9.13}$$

The relationship has been found to hold for many catalysed reactions of this type. When α is small, the solvent is found to be the main catalyst, while when α approaches unity H_3O^+ is the important catalyst corresponding to specific acid catalysis. The above relationships are examples of Hammett or *linear free energy relationships* used to establish the mechanisms of organic reactions.

9.3 Heterogeneous catalysis

Many reactions are catalysed by processes that occur at the boundary between two phases such as a gas–solid or a gas–liquid interface. In these reactions the solid is regarded as the catalyst. Since the rate depends on the concentration of reactants in contact with the surface, it is important that the solid is in a form with a large surface area. A number of industrial processes are good examples of heterogeneous catalysis and show that different catalysts lead to different products in some cases.

(i) Organic decompositions

Ethanol vapour passed over aluminium oxide produces ethylene but with a copper catalyst gives acetaldehyde

$$C_2H_5OH \xrightarrow[300^\circ C]{Al_2O_3} C_2H_4 + H_2O$$

$$C_2H_5OH \xrightarrow[300^\circ C]{Cu} CH_3CHO + H_2$$

(ii) Hydration of unsaturated hydrocarbons

Ethanol vapour is produced from the hydration of ethylene at high pressures using phosphoric acid absorbed on Celite as a catalyst

$$C_2H_4 + H_2O \xrightarrow{300^\circ C} C_2H_5OH$$

(iii) Dehydrogenation

Butane gives butenes, particularly 1,3-butadiene when passed over an aluminium oxide–chromium (III)oxide catalyst

$$CH_3CH_2CH_2CH_3 \rightarrow CH_2=CHCH=CH_2 + 2H_2$$

Ethyl benzene is dehydrogenated to styrene at $650^\circ C$ in the presence of a promoted iron oxide catalyst

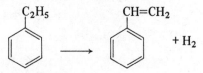

(iv) Hydrochlorination

Vinyl chloride is prepared by reaction of acetylene with HCl gas using a mercuric chloride on charcoal catalyst

$$CH{\equiv}CH + HCl \xrightarrow{200^\circ C} CH_2=CHCl$$

9.3.1 Mechanism of gas–solid reactions

Most of the important industrial catalytic processes occur at a gas–solid interface. The mechanism of these reactions is based on the theory

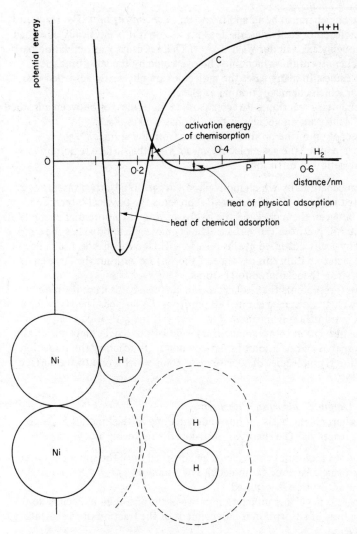

Figure 9.3 *Potential-energy diagram for physical adsorption and chemi-sorption of hydrogen on a nickel surface*

postulated by Langmuir[1] in 1916. He suggested that the process occurred as follows:

(1) Movement of the gas molecules to the surface by convection or diffusion.
(2) Adsorption of reactant molecules on the surface. This adsorption is by

a strong chemical bond and is known as *chemisorption*. The reactant is therefore not easily desorbed, as is the case if it is physically adsorbed through weak van der Waals forces. There is usually an activation barrier to chemisorption, which must be overcome by the adsorbing gas molecules. In many cases the molecules are physically adsorbed first, after which chemisorption takes place.

(3) Elementary reaction between adsorbed molecules, or between adsorbed molecules and molecules in the gas phase, takes place.
(4) Desorption of the product molecules from the surface.
(5) Movement of the gas molecules away from the surface region by convection or diffusion.

The mechanism by which molecular hydrogen is adsorbed on nickel is illustrated in figure 9.3. The curve P represents the physical interaction energy between the nickel and the hydrogen and has a potential energy well at about 340 pm from the surface. The H_2 molecule approaches the surface and is physically adsorbed at this position. This route enables the molecule to approach to within this distance of the surface without the large energy necessary for dissociation into H atoms.

Curve C represents the chemisorption curve and this has a deep potential energy well at a distance about 160 pm from the surface. The H_2 molecules undergo a transition from curve P to curve C if the activation energy barrier B to chemisorption is overcome. The height of this barrier depends on the relative shapes and position of the two curves. On reaching the transition state B the H_2 molecules dissociate into atoms which bond to the surface atoms.

9.3.2 Langmuir adsorption isotherm

This isotherm is the basis for our understanding of the kinetics of heterogeneous reactions. The theory of it makes the following assumptions:

(i) The gas molecules continue to be adsorbed until the surface is covered by a single layer of gas molecules or a monolayer.
(ii) The adsorption is localised.
(iii) There is negligible interaction between the adsorbed molecules so that the heat of adsorption is independent of the fraction of the surface covered.

The isotherm is derived by assuming that a dynamic equilibrium is set up, when the rate of adsorption is equal to the rate of desorption.

Let V be the equilibrium volume of gas adsorbed per unit mass of adsorbent at pressure p, and V_m be the volume of gas required to cover unit mass of adsorbent with a complete monolayer.

The rate of adsorption depends on the following:

(i) The rate of collision of gas molecules with the surface. This is proportional to pressure p.

(ii) The probability of the gas molecule striking a vacant site, that is, part of the surface still free of adsorbed molecules. This probability is equal to $1 - \theta$, where θ is the fraction of the surface covered and is equal to V/V_m.

(iii) The value of the activation energy term $E_{ads}^{\ddagger}$ since the rate of chemisorption is proportional to $\exp(-E_{ads}^{\ddagger}/RT)$.

Combining (i), (ii) and (iii), it is seen that

rate of adsorption $\propto p(1 - \theta) \exp(-E_{ads}^{\ddagger}/RT)$

The rate of desorption depends on:

(i) The fraction of the surface covered, θ

(ii) the activation energy for desorption, $E_d^{\ddagger}$, so that

rate of desorption $\propto \theta \exp(-E_d^{\ddagger}/RT)$

At equilibrium, the rate of adsorption equals the rate of desorption. Therefore

$$p(1 - \theta) \exp(-E_{ads}^{\ddagger}/RT) = k\theta \exp(-E_d^{\ddagger}/RT) \tag{9.14}$$

where k is a constant.

Since the heat of adsorption is equal to $E_{ads}^{\ddagger} - E_d^{\ddagger}$ equation 9.14 rearranges to give

$$p = k\left(\frac{\theta}{1 - \theta}\right)\exp(-\Delta H_{ads}/RT) \tag{9.15}$$

But it was assumed that ΔH_{ads} was independent of the surface covered; that is

$$k \exp(-\Delta H_{ads}/RT) = 1/a$$

where a is a constant depending only on temperature. Equation 9.15 now becomes

$$ap = \frac{\theta}{1 - \theta} = \frac{V/V_m}{1 - V/V_m}$$

or

$$\theta = \frac{V}{V_m} = \frac{ap}{1 + ap} \tag{9.16}$$

This is the Langmuir adsorption isotherm expression. If a plot of p against V is drawn, a limiting value of V is approached at high pressures and this corresponds to the volume occupied by monolayer coverage, V_m.

A set of p-V values can be shown to satisfy equation 9.16 by the following treatment. Inverting equation 9.16 and multiplying throughout by p gives

$$\frac{p}{V} = \frac{1}{aV_m} + \frac{p}{V_m}$$

A plot of p/V against p is linear with a slope of $1/V_m$ and a is given by the slope to intercept ratio.

The Langmuir adsorption isotherm is the basis of our understanding of the rates of heterogeneous reactions. It is found that the order of the reaction depends on the extent of the adsorption.

(i) Reactants weakly adsorbed
The fraction of the surface covered by adsorbed molecules is small; that is, $\theta \ll 1$ so that equation 9.16 becomes

$$ap = \theta$$

The rate can be expressed as the rate of change of pressure with time; therefore

$$-\frac{dp}{dt} = k\theta$$

$$= k'p \tag{9.17}$$

where $k' = ak$. The reaction shows first-order kinetics.

The following well-known catalytic decompositions obey equation 9.17

$$2N_2O \xrightarrow{\text{gold}} 2N_2 + O_2$$

$$2HI \xrightarrow{\text{platinum}} H_2 + I_2$$

$$2PH_3 \xrightarrow{\text{glass}} 2P + 3H_2$$

If in a bimolecular process, both reactants are weakly adsorbed

$$-\frac{dp}{dt} = k\theta^2 = k'p^2$$

and the reaction is second order. Only a few reactions show this behaviour, the bromination of ethylene on the walls of a glass vessel being one example

$$C_2H_4 + Br_2 \xrightarrow{\text{glass}} C_2H_4Br_2$$

(ii) Reactants moderately adsorbed
In this case, the fraction $1 - \theta$ is not negligible, so that the rate equation becomes

$$-\frac{dp}{dt} = k\theta$$

Substituting the expression for θ given in equation 9.16 gives

$$-\frac{dp}{dt} = \frac{kap}{1 + ap} = k'p^n$$

where n is usually less than unity, giving fractional order kinetics. The order is also temperature dependent, since θ changes with temperature.

For the following reaction, $n = 0.6$ at $25^\circ C$

$$2SbH_3 \xrightarrow{\text{antimony}} 2Sb + 3H_2$$

but as the temperature increases θ decreases and n approaches unity. The reactant is only weakly adsorbed under these conditions, as in (i).

(iii) Reactants strongly adsorbed
Under these conditions the number of active sites on the catalyst is few, and the rate is correspondingly low. When the fraction of surface covered approaches unity, the rate becomes independent of the gas pressure so that

$$-\frac{dp}{dt} = k$$

where k is a constant. The reaction therefore shows zero kinetics.

The decomposition of hydrogen iodide, which showed first-order kinetics on platinum is found to obey zero-order kinetics on gold

$$2HI \xrightarrow{\text{gold}} H_2 + I_2$$

The decomposition of ammonia on tungsten is another process that obeys this zero-order expression at pressures above 20 torr pressure.

$$2NH_3 \xrightarrow{\text{tungsten}} N_2 + 3H_2$$

Example 9.1
The following results were obtained for the decomposition of ammonia on a tungsten wire at $856^\circ C$

Total pressure/torr	228	250	273	318
Time/s	200	400	600	1000

Find the order of the reaction and calculate the rate constant.

At high pressures for a strongly adsorbed gas such that $\theta = 1$, and for an increase in total pressure

$$\frac{\mathrm{d}p}{\mathrm{d}t} = k$$

A pressure against time graph is drawn in figure 9.4. Since the graph is linear, $\mathrm{d}p/\mathrm{d}t$ is a constant and the reaction is zero order.

$$\text{slope} = k = 0.113 \text{ torr s}^{-1}$$
$$= 0.113 \times 133.3 \text{ N m}^{-2} \text{s}^{-1}$$
$$= \frac{0.113 \times 133.3}{8.31 \times 1129} \text{ mol m}^{-3} \text{s}^{-1}$$
$$= 1.61 \times 10^{-6} \text{ mol dm}^{-3} \text{s}^{-1}$$

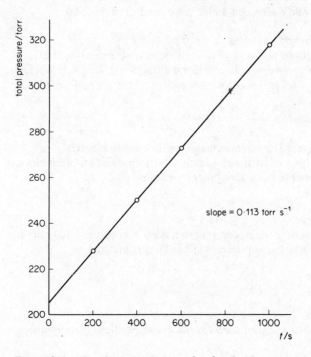

Figure 9.4 *Total pressure–time plot for the decomposition of ammonia on a tungsten wire*

9.4 Enzyme catalysis

Enzymes are biological catalysts that are active in living systems. Since they are proteins with dimensions in the colloidal range and their kinetic behaviour is similar to that shown by heterogeneous catalysts, they are sometimes referred to as microheterogeneous catalysts.

One notable feature is their specificity. Urease is an excellent catalyst for the conversion of urea to ammonia and carbon dioxide

$$CO(NH_2)_2 + H_2O \xrightarrow{\text{urease}} 2NH_3 + CO_2$$

but is not known to catalyse any other processes. There are enzymes that act on optically-active molecules and affect only one of the optical isomers.

Enzyme-catalysed reactions are fast reactions and are studied by some of the techniques described in chapter 11. Kinetic studies on these systems are difficult because pure enzymes are not easily obtained. Further the mechanisms are extremely complex. The enzyme has a number of active sites and often interacts with a substrate in a number of ways. Despite this many reliable rate constants have been determined for enzyme-catalysed reactions. By investigating the effects of concentration, pH, ionic strength and temperature on the rates, an understanding of the mechanism of such reactions has been formulated.

The mechanism of enzyme reactions is thought to proceed as follows. A substrate S catalysed by an enzyme E first complexes with the enzyme to form a substrate–enzyme complex ES, which can either break up to give E and S again or give the products and regenerate E. This mechanism is represented by

$$E + S \underset{k_{-1}}{\overset{k_1}{\rightleftharpoons}} ES \qquad\qquad (1), (-1)$$

$$ES \xrightarrow{k_2} \text{products} + E \qquad\qquad (2)$$

where k_1 is the rate constant for the formation of the complex, k_{-1} is the rate constant for the reverse reaction and k_2 is the rate constant for the dissociation of the complex to yield the products. Michaelis and Menten[2] have derived expressions to explain the effect of substrate concentration on the reaction rate.

Consider the equilibrium between the substrate, enzyme and complex. Let [E] and [S] be the initial concentrations of enzyme and substrate, respectively, and [ES] be the equilibrium concentration of enzyme–substrate complex. The concentration of free enzyme in the system at equilibrium is therefore all of the original enzyme that is not complexed, that is [E] − [ES]. The substrate concentration [S] is usually greatly in excess of [E], so that its equilibrium concentration is essentially equal to its initial concentration [S]. The equilibrium constant K is therefore given by

$$K = \frac{([E] - [ES])[S]}{[ES]}$$

or rearranging

$$[ES] = \frac{[E][S]}{K + [S]}$$

If it is assumed that the rate of reaction (2) is sufficiently slow so as not to disturb the equilibrium, the reaction rate v is therefore given by the rate of reaction (2)

$$v = k_2[ES]$$
$$= \frac{k_2[E][S]}{K + [S]} \tag{9.18}$$

The maximum reaction rate v_{max} or limiting reaction rate is attained when all the enzyme is complexed in the form ES, when the concentration of ES is equal to the initial enzyme concentration [E]. Under these conditions the reaction rate becomes

$$v_{max} = k_2[E] \tag{9.19}$$

Substitution of equation 9.19 into equation 9.18 gives

$$v = \frac{v_{max}[S]}{K + [S]} \tag{9.20}$$

which is known as the *Michaelis equation,* and the equilibrium constant K is known as the *Michaelis constant.*

Lineweaver and Burk[3] showed that the Michaelis equation can be expressed in a linear form from which v_{max} and K can be determined. Taking the reciprocal of each side of equation 9.20 gives

$$\frac{1}{v} = \frac{K + [S]}{v_{max}[S]}$$

that is

$$\frac{1}{v} = \frac{K}{v_{max}[S]} + \frac{1}{v_{max}} \tag{9.21}$$

A plot of $1/v$ against $1/[S]$ is therefore linear with the slope equal to K/v_{max} and the intercept equal to $1/v_{max}$.

Alternatively Eadie[4] suggested that equation 9.20 is best rearranged to give

$$\frac{v}{[S]} = \frac{v_{max}}{K} - \frac{v}{K} \tag{9.22}$$

A plot of $v/[S]$ against v is therefore a straight line of slope $-1/K$. The intercept on the v axis is equal to v_{max}/K. It is claimed that equation 9.22 is less likely to obscure deviations from linearity than equation 9.21.

Example 9.2

Some data for the hydrolysis of methyl hydrocinnamate in the presence of the enzyme chymotrypin is given for the reaction at $25°C$ and pH 7.6

Methyl hydrocinnamate concen-

tration $\times 10^3/\text{mol dm}^{-3}$	30.8	14.6	8.57	4.60	2.24	1.28	0.32
Initial rate $\times 10^8/\text{mol dm}^{-3} \text{ s}^{-1}$	20	17.5	15.0	11.5	7.5	5.0	1.5

Calculate the limiting rate and the Michaelis constant for the reaction.

From equations 9.21 and 9.22 it is seen that both a plot of $1/v$ against $1/[S]$ and a plot of $v/[S]$ against v should be linear

$10^8 v/\text{mol dm}^{-3} \text{ s}^{-1}$	20	17.50	15.00	11.50	7.50	5.00	1.50
$\dfrac{10^{-7}}{v}\Big/\text{dm}^3 \text{ s}^1 \text{mol}^{-1}$	0.500	0.571	0.667	0.870	1.33	2.00	6.67
$\dfrac{10^{-3}}{[S]}\Big/\text{dm}^3 \text{mol}^{-1}$	0.032	0.069	0.117	0.218	0.446	0.781	3.125
$\dfrac{10^{-5} v}{[S]}\Big/\text{s}^{-1}$	0.650	1.20	1.75	2.50	3.35	3.90	4.70

A plot of $1/v$ against $1/[S]$ is given in figure 9.5

$$\text{slope} = K/v_{max} = 1.98 \times 10^4 \text{ s}$$

and

$$\text{intercept} = 1/v_{max} = 0.45 \times 10^7 \text{ dm}^3 \text{ mol}^{-1} \text{ s}$$

Therefore

$$v_{max} = 2.28 \times 10^{-7} \text{ mol dm}^{-3} \text{ s}^{-1}$$

and

$$K = 4.51 \times 10^{-3} \text{ mol dm}^{-3}$$

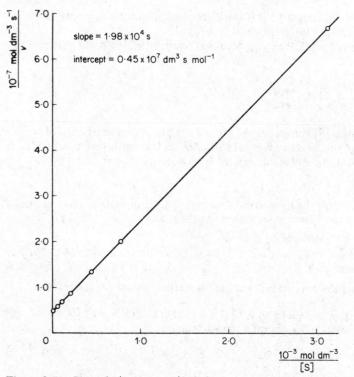

Figure 9.5 *Plot of $1/v$ against $1/[S]$ for enzyme-catalysed hydrolysis of methyl hydrocinnamate*

A plot of $v/[S]$ against v is given in figure 9.6

$$\text{slope} = -1/K = -0.218 \times 10^3 \text{ dm}^3 \text{ mol}^{-1}$$

and

$$\text{intercept} = v_{max}/K = 5.0 \times 10^{-5} \text{ s}^{-1}$$

giving

$$K = 4.6 \times 10^{-3} \text{ mol dm}^{-3}$$

and

$$v_{max} = 2.3 \times 10^{-7} \text{ mol dm}^{-3} \text{ s}^{-1}$$

There is therefore good consistency between the two results, but Eadie claims that his method is less likely to obscure deviations from linearity. A

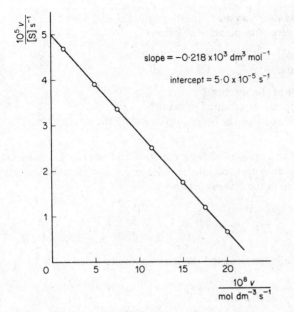

Figure 9.6 *Plot of $v/[S]$ against v for enzyme-catalysed hydrolysis of methyl hydrocinnamate*

comparison of the two curves shows that the Eadie method provides an even distribution of points on the graph, and is therefore to be preferred to the Lineweaver–Burk graph, which depends too much on one reading.

Problems

1. The following data were obtained for the decomposition of glucose at $140°C$ at various concentrations of hydrochloric acid catalyst

$10^4 \, k/\text{min}^{-1}$	6.10	9.67	13.6	17.9
$10^2 [H_3O^+]/\text{mol dm}^{-3}$	1.08	1.97	2.95	3.94

Calculate the catalytic coefficient for H_3O^+.

2. The following results are for the decomposition of diacetone alcohol catalysed by hydroxyl ions at $25°C$

$10^3 [OH^-]/\text{mol dm}^{-3}$	5	10	20	40	100
Rate constant/s^{-1}	3.87	7.78	15.7	32.0	79.9

Calculate the catalytic coefficient for hydroxyl ions.

3. When ammonia decomposes on a tungsten wire, the half-life of the reaction is found to vary with pressure as follows

Pressure/torr	265	130	58	16
Half-life/min	7.6	3.7	1.7	1.0

(a) What is the order of the reaction?
(b) Is the order independent of initial pressure?
(c) What is the significance of the order in considering the mechanism of the reaction?

4. The initial rate of oxidation of sodium succinate to form sodium fumarate in the presence of the enzyme succinate dehydrogenase at different substrate concentrations is given in the following table

Sodium succinate concentration $\times 10^3$/mol dm^{-3}	10.2	2.0	1.0	0.5	0.33
Initial rate/μmol s^{-1}	1.17	0.99	0.79	0.62	0.50

Determine the Michaelis constant and the limiting rate of reaction.

Further reading

References
1. I. Langmuir. *J. Am. chem. Soc.*, **38** (1916) 2221; **40** (1918), 1361.
2. L. Michaelis and M. L. Menten. *Biochem. Z.*, **49** (1913), 333.
3. H. Lineweaver and D. Burk. *J. Am. chem. Soc.*, **56** (1934), 658.
4. G. S. Eadie. *J. biol. Chem.*, **146** (1942), 85.

Reviews
R. P. Bell. Rates of simple acid–base reactions. *Q. Rev. chem. Soc.*, **13** (1959), 169.
D. Shooter. Experimental methods for the study of heterogeneous reactions. *Comprehensive Chemical Kinetics*, **1** (1969), 180.
S. Doonan. Chemistry and physics in enzyme catalysis. *R. Inst. Chem. Rev.*, **2** (1969), 117.
L. Peller and R. A. Alberty. Physical chemical aspects of enzyme reactions. *Progr. Reaction Kinetics*, **1** (1963), 235.

Books
R. P. Bell. *Acid–Base Catalysis*, Clarendon Press, Oxford (1941).
R. P. Bell. *Acids and Bases*, Methuen, London (1952).
P. G. Ashmore. *Catalysis and Inhibition of Chemical Reactions*, Butterworths, London (1963).
S. J. Thomson and G. Webb. *Heterogeneous Catalysis*, Oliver and Boyd, Edinburgh (1968).
K. J. Laidler. *The Chemical Kinetics of Enzyme Action*, Clarendon Press, Oxford (1958).

10 PHOTOCHEMICAL REACTIONS

One of the simplest methods of generating an atom or a free radical is to decompose a suitable molecule by irradiation with light of a suitable wavelength. Provided the molecules can absorb the light energy, they will be excited from their ground state to an excited state, which, in some cases, can undergo homolytic bond scission into atoms or free radicals. Since the number of free radicals produced in a given time will depend on the intensity of the light absorbed, it is very simple to control the experimental conditions of a photochemical reaction. Photochemical techniques provide a convenient method of investigating the kinetics of free-radical reactions at room temperature or low temperatures. It would be necessary to use much higher temperatures to generate free radicals by thermal methods.

While it is possible to induce bond scission by high-energy radiation such as α, β and γ particles, protons, neutrons and X-rays, it is conventional to consider such processes as being radiolytic. The term 'photochemical' is usually confined to processes that are initiated by visible or ultraviolet light. Radiation at these wavelengths induces a transition from the ground electronic state to an excited electronic state.

10.1 Laws of photochemistry

When an absorbing medium is irradiated, the light can be transmitted, scattered, refracted or absorbed. It is the first fundamental law of photochemistry that only light absorbed by the system can be effective in producing a photochemical reaction. For example, an aldehyde such as acetaldehyde can be irradiated with light at 366 nm, but no reaction will occur since acetaldehyde does not absorb radiation at wavelengths greater than 340 nm.

When a beam of light of intensity I_0 and wavelength λ, is passed through a sample of gas or liquid, the light transmitted by the system I_t is found from Beer's law to be given by

$$I_t = I_0 \exp(-\kappa c l)$$

where κ is the molar Naperian extinction coefficient, c is the concentration of the absorbing species and l is the pathlength of the beam through the sample. From the above

$$\ln (I_t/I_0) = -\kappa cl$$

or

$$\log_{10} (I_t/I_0) = -\epsilon cl$$

where ϵ is the molar decadic extinction coefficient.

The light absorbed I_a is given by $I_0 - I_t$, assuming negligible scattering and refraction, and the decadic absorption A is given by

$$A = \log_{10} (I_0/I_t) = \epsilon cl$$

The amount of light energy absorbed by the system therefore depends on the concentration of the reactant, the length of the reaction cell and the molar decadic extinction coefficient at that wavelength.

For a photochemical reaction to occur, there must be an interaction between the radiation and the reactant. From quantum theory the energy, in this case light energy, is quantised; that is, it can be considered as discrete units of energy known as quanta. The energy of a *quantum E* is given by

$$E = h\nu = hc/\lambda$$

where h is Planck's constant, ν is the frequency of the radiation, c is the velocity of light and λ is the wavelength. One mole of quanta is known as an *einstein* of energy, and is given by

$$E = N_A h\nu = N_A hc/\lambda$$

This equation gives the relationship between the energy available and the wavelength of the reaction, as illustrated by table 10.1. If the quanta of

TABLE 10.1   RELATIONSHIP BETWEEN WAVELENGTH
AND ENERGY

Wavelength/nm	Energy/kJ mol^{-1}
600	199
500	239
400	299
300	398
200	597

energy corresponds to the difference in energy between two states of the molecule, the energy is absorbed, and a transition is said to occur between the two states.

The second law of photochemistry is stated by the Stark–Einstein law of photochemical equivalence as: 'Each molecule participating in a photochemical reaction absorbs one quantum of light' or, in other words, a photochemical reaction is a one-quantum process. The quantum yield of a reaction is given by

$$\phi = \frac{\text{No. of molecules of reactant lost or product formed per unit time}}{\text{No. of quanta absorbed per unit time}}$$

$$= \frac{\text{No. of moles of reactant lost or product formed per unit time}}{\text{No. of einsteins absorbed per unit time}}$$

Each molecule can absorb a quantum of energy and produce an excited molecule. If each of these excited molecules yields a molecule of product, the quantum yield is unity. In practice, the excited molecule can undergo, besides a chemical change, other processes that do not involve the breakage of a bond. This means that the quantum yield can often be less than unity. In other reactions if the primary process of light absorption produces free radicals, a chain reaction can be initiated in which a large number of product molecules are formed for each light quantum absorbed. In this case the apparent quantum yield will be very much greater than unity. It is therefore important to note that the Stark–Einstein law applies only to the primary process in which one quantum of light absorbed produces one excited molecule.

Example 10.1
When acetone was photolysed at 56°C with 313 nm radiation for 23 000 s, 5.23×10^{19} molecules were decomposed. If 8.52×10^3 J of radiation were absorbed per second, calculate the quantum yield.

The energy of one quantum $= \dfrac{hc}{\lambda}$

$$= \frac{6.626 \times 10^{-34} \times 2.998 \times 10^8}{313 \times 10^{-9}} \text{ J}$$

$$= 6.345 \times 10^{-19} \text{ J}$$

Therefore, no. of quanta absorbed per second $= \dfrac{8.52 \times 10^{-3}}{6.345 \times 10^{-19}}$

and during the reaction

$$\frac{8.52 \times 10^{-3} \times 2.3 \times 10^4}{6.345 \times 10^{-19}}$$

quanta are absorbed.

Therefore

$$\phi = \frac{\text{No. of molecules decomposed}}{\text{No. of quanta absorbed}}$$

$$= \frac{5.23 \times 10^{19} \times 6.345 \times 10^{-19}}{8.52 \times 10^{-3} \times 2.3 \times 10^{4}}$$

$$= 0.17$$

10.2 Excited-molecule processes

A primary photochemical process is a unimolecular reaction involving a single molecule and a photon. The transition to the excited state occurs in about 10^{-16} s, so that, according to the *Franck–Condon principle*, the molecule undergoes negligible change in bond length due to vibration during the transition. The energy distribution of the excited molecule will be non-Boltzmann, and, compared to the ground state, will contain much more energy. Such energy-rich molecules are termed 'hot' molecules; that is, they are electronically, vibrationally or rotationally 'hot'.

The 'hot' molecule will normally have a life-time of about 10^{-8} s, after which, if no other process has occurred, it will re-emit its energy and revert to a ground-state molecule again. On the other hand the excited species may within that life-time undergo: (a) dissociation; (b) deactivation or chemical reaction by collisions with other molecules; or (c) internal energy changes.

The subsequent fate of the excited molecule depends on the shape and position of the potential-energy curve of the excited state relative to the ground state. This is illustrated in figure 10.1 where four transitions are shown. In figure 10.1(a) the two potential-energy curves are similar in shape and position. This results in the formation of a stable excited state. In figure 10.1(b) the excited state has a greater equilibrium internuclear distance than in the ground state, and dissociation occurs. This will always occur when the energy absorbed exceeds the bond dissociation energy of the excited molecule. In figure 10.1(c) the shape of the excited state shows that it is unstable, and any transition from the ground state will lead to dissociation. In figure 10.1(d) the stable excited state is intersected by an unstable excited state. On excitation the molecule will switch from the stable to the unstable excited state and dissociate, even though the energy absorbed is less than the bond dissociation energy of the stable excited state. This phenomenon is known as *predissociation*.

In a typical organic molecule, there is usually a stable triplet[†] state of

---

† In spectroscopy the multiplicity of a species is given by $2S + 1$, where S is the total electron spin. In most molecules all the electrons are paired and $S = 0$, so that the multiplicity is 1. Such molecules are in the singlet state. A monoradical has one unpaired electron, $S = \frac{1}{2}$, and the multiplicity is 2. These are doublet species. For a molecule or diradical with two unpaired electrons, $S = 1$, and the multiplicity is 3. These species are in the triplet state.

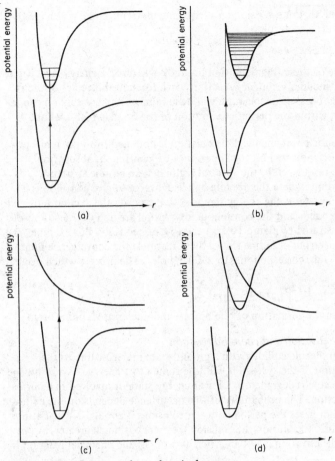

Figure 10.1 *Primary photochemical processes*

lower energy than the lowest excited singlet state. Although transitions between singlet and triplet states are spin-forbidden, energy transfer between them by a radiationless process is possible. Because such processes are spin-forbidden, they occur with a much lower probability than transitions in which no change of spin is involved.

10.2.1 Dissociation

Dissociation of a 'hot' molecule will lead to the formation of two or more atoms or free radicals, and in some cases a molecule. For a diatomic molecule, homolytic bond scission gives two monoradicals

$$AB \xrightarrow{\;h\nu\;} A\cdot + B\cdot$$

while heterolytic bond scission of a triatomic molecule could give a diradical and a neutral molecule

$$AB \xrightarrow{h\nu} A: + B$$

One or more of these fragmentation products will also be energy-rich. These species will undergo secondary reactions with reactant molecules or each other. This process of dissociation will occur almost immediately on excitation; that is, within the period of vibration of the bond, which is about 10^{-14} s.

Evidence for photochemical dissociation is obtained from the molecular spectrum. It is indicated by the appearance of 'continuous absorption', which is the term given to the spectral region where no fine structure occurs. The limit where the discrete bands converge to give continuous absorption represents the energy required to dissociate the excited molecule. The discrete bands and the continuous absorption are due to transitions of the type illustrated by figure 10.1(a) and (b), respectively. For example, the molecular spectrum of iodine shows these features, the convergence limit being at 499 nm, corresponding to 240 kJ mol^{-1}. The process which occurs is

$$I_2 \xrightarrow{h\nu} I\cdot(^2P_{3/2}) + I\cdot(^2P_{1/2})$$

and results in the formation of one normal and one excited iodine atom.

10.2.2   Deactivation and chemical reaction

If the 'hot' molecule collides with a ground-state molecule due to the thermal motion of the system, it will lose some of its energy. As a result the excited molecule is degraded to a lower energy state or involved in a normal thermal reaction. The probability of this occurrence depends on the collision frequency. For gases the period between collisions is about 10^{-10} s at atmospheric pressure. In solution the collision frequency is much greater, but in some cases deactivation is not an efficient process, and chemical reaction is favoured.

10.2.3   Internal energy changes

If, during the lifetime of the excited molecule, it neither dissociates nor undergoes molecular collision, it can revert to a lower energy level by re-emitting its energy. If the same quantum of energy is emitted as was absorbed, this will be the only process, and the light emitted is said to be 'resonance radiation'. For example, when a ground-state mercury atom is excited by 253.7 nm radiation, it will revert to the ground state by emitting the 253.7 nm resonance line. This process occurs in low-pressure mercury-vapour lamps, which are, therefore, an excellent source of monochromatic (253.7 nm) radiation.

It is possible for the excited state to undergo internal energy changes, and by radiationless processes lower its energy in stages to a lower vibrational state of the excited singlet state. If radiation is emitted from this state to any vibrational state of the ground electronic state, it is termed *fluorescence*. The radiation will be of a longer wavelength than the quantum absorbed, and in general consists of a number of lines at different wavelengths corresponding to transitions from and to different vibrational levels. For example rhodamine B will give red fluorescence when it is irradiated with blue, green or red light.

Fluorescence will only be observed in gases at low pressures when the lifetime of the excited state is longer than the time between collisions. For fluorescence to be observed in solution, the excited molecule must be resistant to collisional deactivation. It is found that the intensity of the fluorescence radiation is dependent on the concentration of the solution and the nature of the solvent. Conjugated organic molecules, which have a stable excited state, readily fluoresce.

As has been mentioned already, if a triplet state exists intersystem crossing can occur with a low probability. Triplet states are relatively long-lived ($\approx 10^{-3}$ s) and will often undergo dissociation or deactivation. The molecule is now in a metastable state since reversion to the singlet ground state by emission of radiation is spin-forbidden. However, this process can occur in some cases and is known as *phosphorescence.* Since this metastable state is relatively long-lived, phosphorescence will often continue for a period of from 10^{-3} s to 1 s after the source of radiation is removed.

10.3 Photolytic reactions

Many photochemical reactions involve either dissociation or some subsequent reaction of the photochemically activated molecule. An example of each type of reaction is described to illustrate some of the features outlined in the previous section.

10.3.1 Decomposition of hydrogen iodide

The gas-phase photochemical decomposition of hydrogen iodide into hydrogen and iodine is well understood. It can be represented by a process of the type illustrated in figure 10.1(a), although there are a number of unstable excited states.

Absorption of radiation at long wavelengths corresponds to the primary process

$$HI \xrightarrow{h\nu} H\cdot + I\cdot(^2P_{3/2}) \tag{1}$$

while at some shorter wavelengths, the primary process is

$$HI \xrightarrow{h\nu} H\cdot + I\cdot(^2P_{1/2})$$

The iodine atoms are therefore formed in either the ground state ($^2P_{3/2}$) or the excited ($^2P_{1/2}$) state. The light hydrogen atoms, although they are not electronically excited, are translationally extremely 'hot'.

The experimental quantum yield for the decomposition was found to be 2.0, indicating that two molecules of HI are decomposed for each quantum absorbed. The following secondary reactions are thought to occur

$$H\cdot + HI \overset{k_2}{\to} H_2 + I\cdot \qquad (2)$$

$$I\cdot + I\cdot \to I_2 \qquad (3)$$

A steady-state treatment for H atoms gives

$$d[H\cdot]/dt = I_{abs} - k_2[H\cdot][HI] = 0$$

where I_{abs} is the intensity of light absorbed. Therefore

$$I_{abs} = k_2[H\cdot][HI]$$

and the rate of decomposition of HI is given by

$$-d[HI]/dt = I_{abs} + k_2[H\cdot][HI]$$
$$= 2I_{abs}$$

This reaction scheme is therefore in agreement with the experimental data. The alternative process

$$I\cdot + HI \to H\cdot + I_2$$

is about 146 kJ mol^{-1} endothermic, and is unlikely to be favoured over reaction (2), which is 134 kJ mol^{-1} exothermic.

The experimental quantum yield is constant over a wide range of wavelengths and at pressures from 10^{-2} torr to 1 atm. Since no fluorescence is observed even at the lowest pressures, it is unlikely that any excited HI molecules are involved in the mechanism. This is a good example of a photochemical dissociation process.

10.3.2   Dimerisation of anthracene

When a solution of anthracene in benzene is irradiated with ultraviolet light, dimerisation occurs and dianthracene is formed. In dilute solution fluorescence is observed, but its intensity decreases with increasing concentration of anthracene. The quantum yield of dimer, however, increases as the concentration increases. These experimental observations are consistent with the mechanism

$$A \xrightarrow{h\nu} A^* \qquad \text{Activation}$$
$$A^* \longrightarrow A + h\nu' \qquad \text{Fluorescence}$$
$$A^* + A \longrightarrow A_2 \qquad \text{Dimerisation}$$

As the concentration increases, the activated molecule is more and more likely to lose its excess energy by collision with a ground-state molecule to form the dimer than to emit it as fluorescence radiation.

The experimental quantum yield in terms of dimer formed is much less than the theoretical value of unity predicted by this simplified mechanism. Other processes such as deactivation by collision with solvent molecules and thermal dissociation of the dimer also occur.

10.4   Photosensitised reactions

Some molecules that do not absorb in a convenient wavelength range can be dissociated in the presence of another atom or molecule that does absorb the radiation. Such a process is known as *photosensitisation*. The role of the sensitiser is to absorb the light energy and transfer it by collision to the reactant.

Metal vapours are very useful photosensitisers. They must have a high vapour pressure at reasonably low temperatures and have an excitation energy of the right order of magnitude. The most common photosensitiser is mercury vapour. Irradiation of a system containing mercury vapour will produce the following transitions

$$Hg(6^1S_0) + h\nu \ (184.9 \ nm) \rightarrow Hg(6^1P_0)$$

$$Hg(6^1S_0) + h\nu \ (253.7 \ nm) \rightarrow Hg(6^3P_0)$$

which correspond to 645 kJ mol^{-1} and 470 kJ mol^{-1}, respectively. Irradiation of a system containing mercury vapour with a low-pressure mercury-vapour lamp (which emits these two lines as resonance radiation) excites the Hg atom. On collision with the reactant 470 kJ mol^{-1} of energy is available to dissociate the reactant. The 184.9 nm line is absorbed completely by 15 cm of air, but the above energy from the 253.7 nm line is sufficient to dissociate many molecules. For example, nitrous oxide under these conditions dissociates by

$$Hg(6^3P_0) + N_2O \rightarrow Hg(6^1S_0) + N_2 + O(^3P)$$

This process is known as *mercury-photosensitised decomposition,* and is an excellent method of producing oxygen atoms. Since the excited mercury atom is a triplet species, by the spin conservation rule the oxygen atom will be in its triplet ground state. Any molecule that has a bond weaker than 470 kJ mol^{-1} and does not absorb at 253.7 nm can undergo mercury-photosensitised decomposition, for example hydrocarbons, ammonia, hydrogen

$$RH + Hg^* \rightarrow R + H + Hg$$

$$NH_3 + Hg^* \rightarrow NH_2 + H + Hg$$

$$H_2 + Hg^* \rightarrow H + H + Hg$$

In organic systems it is found that aromatic ketones are very useful as photosensitizers. On excitation these compounds readily undergo inter-system crossing and are formed in their triplet state. A reactant that will not absorb the energy itself can be formed in the triplet state by a triplet–singlet energy transfer process. For example, benzophenone will absorb radiation and, by very efficient intersystem crossing, form the biradical triplet state. This is indicated by the absence of fluorescence and the marked effect of oxygen, which scavenges triplet species. Benzophenone donates its triplet energy (286 kJ mol^{-1}) to an acceptor molecule A, which is excited to its triplet state and undergoes dissociation or further reaction.

$$C_6H_5COC_6H_5 \xrightarrow{h\nu} (C_6H_5COC_6H_5)_s{}^*$$

$$(C_6H_5COC_6H_5)_s{}^* \longrightarrow (C_6H_5COC_6H_5)_t{}^*$$

$$(C_6H_5COC_6H_5)_t{}^* + A \longrightarrow C_6H_5COC_6H_5 + A_t{}^*$$

10.5   Experimental methods

It is relatively simple to set up an apparatus for preparing compounds by photochemical methods. The main requirement is a powerful source of radiation of a suitable wavelength set in a position where the maximum amount of radiant energy can be absorbed. A typical photochemical reactor manufactured by Hanovia Instruments Ltd, is shown in figure 10.2.

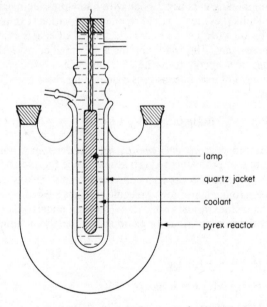

lamp

quartz jacket

coolant

pyrex reactor

Figure 10.2   *A typical photochemical reactor for liquid-phase reactions*

For more quantitative work where spectroscopic data, rate data or quantum-yield measurements are required, it is necessary to use apparatus of the type illustrated by figure 10.3, normally set up on an optical bench. This involves a suitable source of radiation and the light is collimated to pass a parallel beam through the reaction cell perpendicular to the front wall of the vessel. The cell is made of glass for wavelengths down to 340 nm,

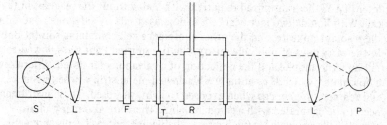

Figure 10.3 *Apparatus for a typical photochemical experiment thermostat*

S = light source; L = lenses; F = filter; T = air or water thermostat; R = reaction vessel; P = photodetector or chemical actinometer.

or of quartz for lower wavelengths. The vessel can be kept in a water thermostat for experiments at wavelengths in the visible range, but, since water itself absorbs u.v. light, it is necessary to use an air thermostat in the u.v. wavelength range. If the quantum yield is to be determined, a detector to measure the quantity of light absorbed during the photolysis can be used.

10.5.1 Light sources

Normally, a metal-vapour lamp such as a mercury-vapour lamp is used. A low-pressure (about 10^{-4} torr) mercury-vapour lamp emits two resonance lines at 253.7 nm and 184.9 nm. If an air gap of greater than 15 cm is used, the 184.9 nm radiation is absorbed by the atmosphere, and these lamps prove to be a valuable source of monochromatic radiation at 253.7 nm suitable for the study of mercury-photosensitised reactions.

Medium-pressure (about 1 atm) mercury-vapour lamps are more powerful because they emit over a wide range of wavelengths in the visible and u.v. range. Two of the strongest bands are centred at 313 nm and 366 nm, and these lamps are very useful for the study of direct photolysis reactions. Other metal-vapour lamps (for example, zinc and cadmium) are less powerful, but often emit at a required wavelength for a certain experiment. For wavelengths less than 200 nm, the emission from low-pressure discharges in hydrogen, krypton and xenon are widely used. More recently lasers have also been adapted as a source of electromagnetic radiation in the u.v., visible and i.r. regions.

In many photochemical reactions it is often desirable to limit the radiation to a suitable wavelength band. This is achieved by the use of a monochromator, an interference or transmission filter, all commercially available, or by a number of chemical solution filters readily prepared in the laboratory.

10.5.2 Chemical actinometers

In order to determine the quantum yield, it is necessary to measure the intensity of the transmitted radiation with and without the reactant in the cell. With liquids two identical cells can be used side by side, one containing the reactant mixture, and the other a reference cell containing solvent only. In the early days of photochemistry, the transmitted light was allowed to fall on a thermopile and the deflection of a galvanometer noted. Alternatively a bolometer, which is essentially a blackened metal strip whose resistance and temperature change when exposed to light, was used. These instruments need to be calibrated with a standard lamp. There are also many photoelectric cells available that will give a reliable light intensity measurement.

However, the most convenient and widely used method of measuring the total light absorbed is termed *chemical actinometry*. The basis of the method is to measure the change produced in a well-defined photochemical reaction under identical experimental conditions. This reaction, called the chemical actinometer, must be simple, reproducible and have a known response over a wide range of wavelengths.

The most widely used system is the uranyl oxalate decomposition. If oxalic acid is exposed to u.v. light in the presence of uranyl ions, the latter catalyses the photodecomposition

$$UO_2^{2+} + H_2C_2O_4 \xrightarrow{h\nu} UO_2^{2+} + CO_2 + CO + H_2O$$

The quantum yield for this reaction is about 0.58 at a number of wavelengths and the number of moles of oxalic acid decomposed per unit time can be measured by titration against standard potassium permanganate solution.

Alternatively the potassium ferrioxalate actinometer is used, particularly in the visible region

$$K_3Fe(C_2O_4)_3 \xrightarrow{h\nu} Fe^{2+} + CO_2, \text{etc.}$$

The Fe^{2+} ions formed can be complexed with 1,10-phenanthroline and measured photometrically. This reaction has a quantum yield that changes slowly with wavelength.

In gas reactions the photolysis of acetone has been shown to have a quantum yield of approximately unity at a number of wavelengths under certain conditions of temperature and pressure. In many gas reactions an actinometer such as the acetone photolysis is not necessary. Many sources of free radicals used in photochemical free-radical reaction investigations yield an inert gaseous product in addition to the free radicals. If the reaction of methyl radicals with an alkane is being studied, a convenient source of methyl radicals is the photodecomposition of azomethane at 366 nm

$$(CH_3)_2N_2 \xrightarrow{h\nu} 2CH_3\cdot + N_2$$

The yield of nitrogen is a measure of the concentration of methyl radicals produced, since one molecule of nitrogen is formed for each pair of methyl radicals generated. The measurement of the yield of nitrogen is therefore an internal actinometer, that is, an inbuilt measure of the number of methyl radicals formed per unit time. In many kinetic studies this is sufficient to enable the relative rates of radical attack on a reactant molecule to be measured.

Example 10.2

A uranyl oxalate actinometer was irradiated with ultraviolet light for 3 hours, during which time 8.6 millimoles of oxalate were decomposed. If the quantum yield at the wavelength used is 0.57, calculate the intensity of the light used in quanta per second.

$$\phi = \frac{\text{No. of molecules of oxalate decomposed}}{\text{No. of quanta absorbed}}$$

that is

$$0.57 = \frac{0.0086 \times 6.02 \times 10^{23}}{\text{No. of quanta absorbed in 3 hours}}$$

Therefore

$$\text{No. of quanta absorbed per second} = \frac{0.0086 \times 6.023 \times 10^{23}}{3 \times 3600 \times 0.57}$$
$$= 8.41 \times 10^{17} \text{ quanta s}^{-1}$$

Problems

1. In the photochemical reaction between H_2 and Br_2, 3.2×10^{17} photons are absorbed per second. After 1200 s irradiation analysis showed that 6×10^{-6} mole of HBr had been formed. What is the quantum yield for the formation of HBr?

[University of Southampton BSc (Part 1) 1972]

2. The quantum yield for the photochemical decomposition of hydrogen iodide is two. How many molecules of hydrogen iodide per kJ of radiant energy absorbed will be decomposed by ultraviolet radiation at 253.7 nm?

3. A gas at $12.53\ \mathrm{N\,m^{-2}}$ and $83^{\circ}\mathrm{C}$ is irradiated for 20.5 h. If 9.95×10^{10} quanta of light are absorbed per second, calculate the number of moles of gas decomposed if the quantum yield is unity.

4. The mechanism for the photochemical reaction between hydrogen and iodine vapour at 480 K is believed to be

$$I_2 + h\nu \xrightarrow{k_1} 2I\cdot$$
$$2I\cdot + I_2 \xrightarrow{k_2} 2I_2$$
$$2I\cdot + H_2 \xrightarrow{k_3} I_2 + H_2$$
$$2I\cdot + H_2 \xrightarrow{k_4} 2HI$$

Show that

$$\frac{d[HI]}{dt} = \frac{2I_{abs}k_4[H_2]}{k_2[I_2] + k_3[H_2]}$$

provided that $k_4 \ll k_3$ and I_{abs} is the intensity of light absorbed.

5. Assuming that the photochemical mechanism for the hydrogen–bromine reaction is essentially the same in the thermal reaction (see page 83) except that the initiation step is

$$Br_2 \xrightarrow{h\nu} Br\cdot + Br\cdot$$

derive an expression for the rate of formation of hydrogen bromide.

Further reading

Reviews

K. P. Suppan. Photochemistry, some recent advances and problems. *Chem. Brit.*, **4** (1968), 538.
J. N. Pitts, Jr, F. Wilkinson and G. S. Hammond. The vocabulary of photochemistry. *Advances in Photochemistry*, **1** (1963), 1.

Books

J. C. Calvert and J. N. Pitts. *Photochemistry*, Wiley, New York (1966).
A. Cox and T. J. Kemp. *Introduction to Photochemistry*, McGraw-Hill, London (1971).
R. B. Cundall and A. Gilbert. *Photochemistry*, Nelson, London (1970).
R. B. Wayne. *Photochemistry*, Butterworths, London (1970).

11 FAST REACTIONS

The field of fast reactions has expanded in recent years with the increase in the use of automatic and electronic recording equipment. The term fast reaction is best applied to reactions that cannot be followed kinetically by conventional methods. It is not sufficient to say that a reaction with a large rate constant is a fast reaction, since this term is temperature dependent. It is more accurate to assume that a fast reaction will have a low activation energy, but if the reactants are present in low concentrations the rate of the reaction will be small. Similarly in a unimolecular reaction the rate of decomposition may be very fast, but if deactivation is effective the rate of conversion to product is small.

A fast reaction is therefore best defined as one whose half-life is less than a few seconds (that is, of the same order as the human response time or the mixing time for reactants) at room temperature using conventional reactant concentrations (say $0.1 \, mol \, dm^{-3}$).

A number of fast gas-phase reactions have been studied by techniques described in this chapter, particularly flow methods. They invariably involve free-radical reactions. Modern analytical instruments, fast and sensitive enough to detect free-radical species that exist for only times of the order of milliseconds are used.

Many more liquid-phase fast reactions have been studied, particularly reactions involving ions or electrons in aqueous solution; that is, hydrated ions or electrons. The fastest reaction studied is the neutralisation reaction

$$H^+ + OH^- \rightarrow H_2O$$

which has a rate constant of $1.4 \times 10^{11} \, dm^3 \, mol^{-1} \, s^{-1}$ at $25\,^\circ C$. Many reactions of biological importance such as enzyme-catalysed reactions are very rapid.

Table 11.1 shows the range of half-lives of reactions that can be studied by techniques outlined in this chapter.

One possible method of investigating a fast reaction is to carry out the reaction under different experimental conditions in which the reaction proceeds at a measurable rate. For instance, a reaction with an appreciable activation energy (for example, $100 \, kJ \, mol^{-1}$) proceeds about 10^8 times slower if the temperature is dropped from $300 \, K$ to $200 \, K$. Alternatively if

TABLE 11.1 RANGE OF HALF-LIVES MEASURABLE BY FAST REACTION TECHNIQUES

Technique	Half-life/s
Conventional	10^3–1
Constant and stopped-flow	1–10^{-3}
Flash and pulse radiolysis	1–10^{-6}
Nuclear magnetic resonance	1–10^{-5}
Electron spin resonance	10^{-4}–10^{-9}
Pressure jump	1–10^{-5}
Temperature jump	1–10^{-6}
Fluorescence	10^{-6}–10^{-9}

the reactant concentrations for a bimolecular reaction are reduced from $0.1 \, mol \, dm^{-3}$ to $10^{-6} \, mol \, dm^{-3}$, the rate will be 10^{10} times slower. However it could be that the data obtained at low temperature or in very diluted solution is not of interest, and in many cases the reaction mechanism could be entirely different. This indirect approach is usually not satisfactory.

Another method that circumvents the need for a fast analytical method is possible when the reaction has a large equilibrium constant

$$A \underset{k_{-1}}{\overset{k_1}{\rightleftharpoons}} B$$

Therefore

$$k_1 = k_{-1} K$$

In this case k_{-1} will be measurable, so that provided K is known the forward rate constant can be determined.

There are traditionally two ways of approach to the consideration of fast reactions. The first involves the division of techniques which are termed:

(i) Perturbation method

A system in equilibrium is subjected to a disturbance or perturbation, and the re-equilibration reaction occurs very rapidly.

(ii) Competition method

A physical process disturbs the system and competes with the chemical reaction. For example, in a photochemical reaction fluorescence is a physical process that competes with chemical processes.

A simpler approach is to consider the different fast reaction experimental techniques used. These have been devised to enable reactions to be initiated very quickly and the subsequent chemical reaction followed analytically. The following methods will be considered:

(1) flow methods;
(2) flames;
(3) flash photolysis and pulse radiolysis;
(4) magnetic resonance methods;
(5) shock tubes;
(6) molecular beams;
(7) relaxation methods.

11.1 Flow methods

This was the first fast-reaction technique invented and is still an important method today. The first apparatus was designed by Hartridge and Roughton in 1923, and is shown in figure 11.1. It consisted of a mixing chamber from which the reactants flowed down a tube at high speed. In most cases the mixture is analysed by absorption spectrophotometry.

The principle of the method is that the concentration of reactant is measured as a function of distance along the flow tube, by measuring the absorption or some other physical property such as the conductivity at

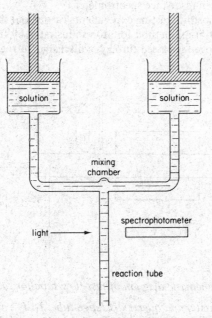

Figure 11.1   *Hartridge and Roughton's constant-flow apparatus*

different distances down the tube. The rate of change of concentration of reactant A with distance is related to the rate of reaction by

$$\frac{-d[A]}{dx} = \frac{-d[A]}{dt} \bigg/ \frac{dx}{dt}$$

where dx/dt is the flow rate. If the flow rate is 10 m s^{-1}, and the observation made at 1 cm (10^{-2} m) distance from the mixing, then this distance is equivalent to a reaction time of $(10^{-2}/10) \text{ s} = 10^{-3} \text{ s}$. In this way it is possible to study reactions with half-lives of the order of milliseconds.

11.1.1 Gas reactions in flow tubes

Many gas reactions involving atoms or free radicals have been studied in flow systems. The gases are passed at relatively low pressures (a few torr) at flow velocities of about 10^3 cm s^{-1} through an electric discharge. The reactions of the free radicals produced are studied at distances up to 1 m downstream from the discharge. In this way it is possible to study free-radical recombination reactions, or the reaction of free radicals with a stable reactant molecule added to the gas stream. Many analytical techniques have been developed, such as emission and absorption spectroscopy, or chemiluminescence, or the gas stream is leaked into a mass spectrometer or an electron spin resonance spectrometer.

Among the most interesting experiments carried out in this field is that of a gas-phase titration method for atoms illustrated by figure 11.2. When oxygen-free nitrogen is passed through a discharge, nitrogen atoms are

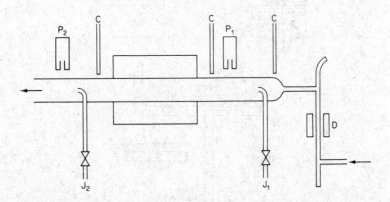

Figure 11.2 *Schematic diagram of fast-flow tube for gas titration*

$D =$ *discharge exciter and quartz discharge tube;* $J_1, J_2 =$ *jets for added gases;* $P_1, P_2 =$ *photomultipliers;* $F =$ *furnace;* $C =$ *cooling air jets*

obtained. The atoms recombine and a straw-yellow coloured emission indicates that excited nitrogen has been formed

$$N \cdot + N \cdot \rightarrow N_2^*$$

A gas-phase titration to estimate the concentration of N atoms is undertaken by adding nitric oxide downstream, when oxygen atoms are formed by

$$N \cdot + NO \rightarrow N_2 + O:$$

As more nitric oxide is added, a blue chemiluminescence is observed from the combination of oxygen atoms with N atoms

$$N \cdot + O: + M \rightarrow NO^* + M$$

where M represents an inert gas—undissociated nitrogen in this case.

At the end point of the titration the nitric oxide has scavenged all of the N atoms and the emission ceases. If nitric oxide is added beyond the end point, a yellow-green air afterglow from excited nitrogen dioxide occurs due to the reaction

$$O: + NO \rightarrow NO_2^* + h\nu$$

11.1.2   Flow reactors for liquid-phase reactions
Two types of flow reactors are used to study reactions in the liquid phase.

(i) Stirred reactor
The reactants are added to a large stirred reaction vessel and the products are removed from the vessel at the same flow rate. The analysis takes place within the reaction vessel itself or alternatively, the products in the exit stream are analysed.

(ii) Stopped-flow reactor
The two reactants are mixed and proceed down the reactor flow tube as in figure 11.3. The flow is suddenly halted by releasing a piston which at the same time triggers a high-speed analytical device. The reaction at a fixed point along the tube is followed.

11.1.3   Limitations of flow methods
There are a number of limitations associated with flow methods.

(i) Hydrodynamic or gas dynamic properties
At high mixing speeds or flow speeds, viscous losses occur and turbulent flow results. There is therefore an upper limit to the flow speed that can be used, and reactions with half lives of less than 10^{-3} s cannot be studied in flow systems.

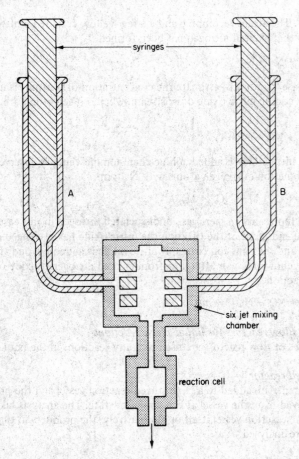

Figure 11.3 *Stopped-flow apparatus*

(ii) Pressure of the gas

In gas reactions in a flow tube low pressures are used, but the pressures must be high enough to avoid changes in concentration occurring by diffusion rather than chemical reaction.

(iii) Volume of reactant

A large volume of reactant is necessary. In early liquid-flow experiments with 5-mm flow tubes, 3–4 dm^3 of reactants were consumed per run. The use of narrow-bore tubing has reduced the volume required to 20–30 cm^3 per run. For very expensive reactants, the use of a mechanically-operated hypodermic syringe has reduced reactant volumes to 6 cm^3 per run.

11.2   Flames

One of the earliest fast-reaction techniques developed was the study of a chemical reaction in a stationary flame. If the reactants are allowed to mix by diffusion within the flame, the reaction is termed a *diffusion flame process*. Alternatively, if the reactants are mixed prior to entry into the flame, a *pre-mixed flame reaction* is obtained.

The pressure of the gases fed into the flame determines the nature and temperature of the flame.

11.2.1   Dilute flames

At low pressures the mean free path between molecular collisions is long and the reaction zone is large. The resultant temperature rise is small and such flames are termed dilute flames.

Much of the early work was concentrated on reactions between alkali metals and halogens or alkyl halides. An inert gas was passed over heated alkali metal, and the resulting gas saturated with metal vapour was diffused into a stream of halogen or halide vapour at low pressures. Reactions of the type

$$RCl + Na \rightarrow R + Na^+Cl^-$$

and

$$Cl_2 + Na \rightarrow Cl + Na^+Cl^-$$

occur in the dilute flame. Provided diffusion coefficients are known, the rate constant can be determined from a measurement of the dimensions of the flame reaction zone and the reactant pressures. Unfortunately the values of diffusion coefficients under flame conditions are not too well established, so that rate constants measured by this technique are not too accurate. The formation of solid product particles in the flame causes further uncertainties.

11.2.2   Hot flames

At high pressures the mean free path is short and the reaction zone is small. When exothermic reactions occur in the flame, there is a rapid rise in temperature. Very sharp temperature gradients occur within the flame, and a temperature of 3000 K is easily achieved.

The emission spectrum of many free-radical species such as OH, NH, NH_2, etc., were first observed in hot flames. These flames result in non-equilibrium conditions in which the electronic, vibrational and translational temperatures are different. The resulting chemiluminescence or chemi-ionisation has been effectively studied in these flames.

11.3   Flash photolysis and pulse radiolysis

Flash photolysis is a modern technique developed by Norrish and Porter by which radicals are produced at relatively high concentrations. Although the lifetimes of many radicals are only of the order of milliseconds or less, they have been identified in flash-photolysis systems by light-absorption methods.

The apparatus is illustrated in figure 11.4. The reactant is decomposed by a high-intensity light flash with energy up to 10^5 J produced by discharging a bank of condensers. The flash, which has a duration of about $100\,\mu s$, is linked by a delay unit to a spectroflash lined up perpendicular to the photoflash. The spectroflash is set to fire at time intervals of the order of $200\,\mu s$ after the photoflash and the absorption spectrum is recorded on the photographic plate of the spectroscope.

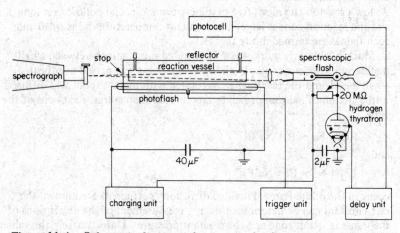

Figure 11.4 *Schematic diagram of apparatus for flash photolysis and kinetic spectroscopy*

Much of the light energy is converted into translational energy, which appears as kinetic energy of the radicals and results in a sudden rise in temperature. If however the reaction is carried out in the liquid phase or with a gaseous reactant in the presence of an inert gas, approximately isothermal conditions pervade. With this technique the absorption spectra of species previously observed only by their emission spectra were detected for the first time. In this way the bond distances and moments of inertia of such radicals as NH_2, C_3, CHO and ClO were determined. In addition the triplet species of many polycyclic molecules in the gas and liquid phase have been observed. An investigation of the rate of the change in the concentrations of these intermediates formed by flash photolysis is termed *kinetic spectroscopy*. This technique has been used to study the kinetics of many fast photochemical decomposition and oxidation reactions.

In kinetic spectroscopy the spectroscopic and kinetic record of the reaction is obtained by taking a series of photographs of the spectra at different time intervals, each photograph requiring a separate experiment. If the spectrograph is replaced by a photomultiplier, and the emission from a particular species monitored by the photomultiplier to give an output signal on an oscilloscope screen, the technique is termed *kinetic spectrometry*. This has proved to be a useful method of studying a number of radical–radical and radical–ion reactions.

For example, the flash photolysis of iodine has been used[1] to measure the rate of recombination of the resultant iodine atoms

$$I\cdot + I\cdot + M \rightarrow I_2 + M^* \tag{1}$$

The recombination has been studied in the presence of a number of third body molecules, that is, where M is H_2, Ar, Ne and CH_4, and the competition of reaction (1) with

$$I\cdot + I\cdot + I_2 \rightarrow I_2 + I_2^* \tag{2}$$

has also been studied. The concentration of iodine molecules can be measured by monitoring the intensity of a narrow beam of light around 500 nm, and the concentration of iodine atoms determined from the expression

$$[I\cdot]_t = 2([I_2]_0 - [I_2]_t)$$

where subscript t refers to time t and subscript 0 to the pre-flash conditions. From this measurement the relative rates of reactions (1) and (2) can be obtained for a number of gases.

An analogous technique is *pulse radiolysis* where the main difference is that ionising radiation is used instead of the photolytic flash. A pulse of electrons from a linear accelerator or X-rays are passed through a solution and the resultant reactions between ions, electrons, molecules, etc., form the interesting field of radiation chemistry.

11.4 Magnetic resonance methods

Nuclear magnetic resonance and electron paramagnetic resonance spectroscopy have both been utilised to measure rapid exchange reactions. Graphs of absorption against frequency known as line shapes provide information about energy transfer in molecular encounters. The structure and width of the line shapes change when a rapid exchange reaction occurs.

Consider the simple n.m.r. spectrum of pure dry ethanol. The spectrum as shown in figure 11.5(a) includes the hydroxyl triplet and the methylene modified quadruplet. The addition of a trace of alkali is detected by the appearance of the water proton line between the peaks. As alkali is added a proton transfer between the OH group and the other species like OH^- and H_2O occurs. This results in the sharpening of the OH triplet to a single peak

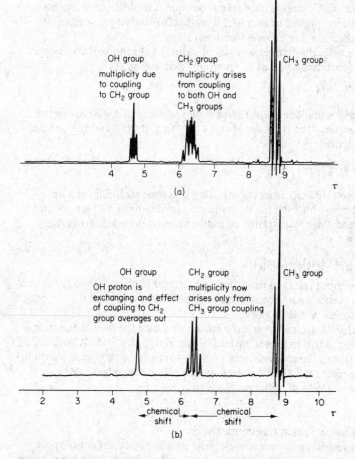

Figure 11.5 *n.m.r. spectrum of pure ethanol vapour (a) The high-resolution n.m.r. spectrum of ethyl alcohol showing the chemical shifts of the different types of proton and the fine structure due to spin–spin coupling. (b) In the presence of alkali the OH proton exchanges so coupling to CH₂ group is averaged*

after which it broadens, and the CH₂ signal changes into a simple quadruplet. The rate of these spectral changes is therefore a measure of the rate of proton exchange. Other changes to the spectrum can be brought about by the addition of acid or base, which causes the hydroxyl proton to exchange in an alcohol such as ethanol

$$ROH + A \rightarrow RO^- + AH^+$$

With proton n.m.r. experiments, mean life times of between 1 and 10^{-4} s can be measured. The rate constant for the reaction

$$CH_3OH + CH_3O^- \rightarrow CH_3O^- + CH_3OH$$

was measured as 8.8×10^{10} s^{-1} in pure methanol.

Electron transfer reactions with life times as low as 10^{-4} to 10^{-9} s can be studied by electron paramagnetic resonance methods. A rate constant of 5×10^8 dm^3 mol^{-1} s^{-1} for the reaction

$$C_6H_5CN + C_6H_5CN^- \rightarrow C_6H_5CN^- + C_6H_5CN$$

has been measured.

11.5 Shock tubes

The shock tube has been developed in the last 20 years as an excellent method of studying fast homogeneous gas-phase reactions at high temperatures. A reactant gas present at low concentrations in excess inert gas undergoes adiabatic compression in a shock wave and is heated to temperatures up to 5000 K or more behind the shock front. A typical shock tube is illustrated in figure 11.6. It consists of a driver section containing the driver gas—hydrogen or helium—a diaphragm and an experimental section for the

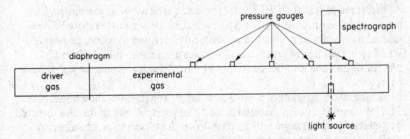

Figure 11.6 *Schematic diagram of a shock tube*

gas under investigation. Gauges are placed along the tube to measure the speed of the shock wave. On bursting the diaphragm, the shock-heated gas is swept down the tube at a speed very much greater than that of sound. The shock front represents a very sharp pressure and temperature, high-resolution profile. Consequently temperatures up to 5000 K are reached within 1 μs.

Reactions can be followed by measuring density changes, light emission or light absorption, the latter technique requiring a wide shock tube. Even higher temperatures can be attained when the shock tube is allowed to reflect at the end of the shock tube. Bradley[2] has studied reactions behind the shock wave reflected at a piece of gold foil. He analysed the reaction mixture that leaks from a pin hole in the foil into a time of flight mass spectrometer, which gave a complete mass spectrum of the species present every $50\,\mu s$.

A large number of decomposition and autoxidation reactions have been studied at temperatures that cannot readily be attained by any other technique. Also the temperature and pressure are accurately defined by the hydrodynamic properties of the shock wave and the thermodynamic properties of the gas. The reactions studied are truly homogeneous since the observation time is short compared to the time necessary for the molecules to diffuse to the walls.

On the other hand a shock tube is an elaborate piece of apparatus to set up and it is often necessary to use expensive electronic equipment. There are gas-dynamic limitations to the experimental conditions that can be used. Attenuation or deceleration of the shock wave occurs as it travels down the tube. This is due to a gradual build-up of a boundary layer between the shock-heated gas and the cold walls of the shock tube. Corrections for this behaviour can be made, but problems of this type usually necessitate the use of a computer in the calculation of rate constants.

11.6   Molecular beams

At pressures of less than 10^{-5} torr, the mean free paths of gas molecules become of the order of metres. The molecular flow can therefore be considered as a beam of non-colliding molecules, termed a *molecular beam*. When two such beams cross, the molecules are scattered in different directions and the number of scattered molecules is measured by a suitable detector.

A schematic diagram of a molecular beam apparatus is shown in figure 11.7. A source of beam material, usually an oven, allows molecules directed by a suitable slit arrangement to pass down the observation chamber. A similar source at right angles provides a second beam. The velocity of the beam is determined from the temperature of the source. The detector slit and detector are moveable about a circle and centre on the point where the beams cross. In this way it is possible to study the intensity of scattering as a function of the angle of scattering and the initial velocity of the beams. Studies of this type have shown that in bimolecular collisions there is:

(a) elastic scattering—a collision that results in no transfer of energy;
(b) inelastic scattering—a collision in which energy is transferred; and
(c) reactive scattering— a collision that produces a chemical reaction.

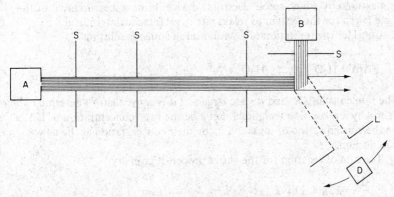

Figure 11.7 *Schematic diagram of molecular beam apparatus*
A = oven; B = source of second beam; S = collimation slits; L = detector slit;
D = movable detector

A large number of the reactions are of the type studied in flames
mentioned in section 11.2. In general they are of the form

$$RX + M \rightarrow R + MX$$

where R is an alkyl, halogen or hydrogen atom, X is a halogen atom and M
is an alkali metal. Their *reactive cross-sections*, a measure of the probability
that reaction will occur when two molecules collide, are orders of magnitude
greater than for most other bimolecular reactions and the corresponding
ions or the reactants themselves can be detected easily. Alkali metals and
metal halides have low vapour pressures so that any molecule not detected
condenses on the vessel walls.

These experiments have shown that in reactions between species of small
reactive cross-section (about 0.1 nm^2) the product molecule MX leaves the
encounter backwards; that is, in the direction from which M approached.
For reactions between molecules of large cross-sections (about 1nm^2), MX
leaves the encounter in the direction from which RX approached. Although
it is unwise to draw general conclusions for all bimolecular reactions from
these results, these experiments are a great step forward in the understanding
of the dynamics of elementary chemical processes.

11.7 Relaxation methods

A number of the techniques described in this chapter have been used to
study chemical relaxation. The system is allowed to come to equilibrium.
It is then disturbed by a sudden impulse (for example, a sudden temperature
and pressure impulse from a shock wave) and the system is no longer at
equilibrium. The speed at which the reaction approaches a new equilibrium

is measured by a high-speed electronic device. From a measurement of the time taken for the system to relax, rate constants are determined.

Consider the ionisation of a weak acid in aqueous solution

$$HA + H_2O \underset{k_{-1}}{\overset{k_1}{\rightleftharpoons}} H_3O^+ + A^-$$

The rate constants k_1 and k_{-1} are large and it is convenient to determine their values by a relaxation technique. Let a be the total concentration of HA, x be the concentration of ions, and x_e be their concentration at the new equilibrium.

The rate of reaction for the above process is given by

$$\frac{dx}{dt} = k_1(a - x) - k_{-1}x^2$$

At equilibrium $dx/dt = 0$ therefore

$$k_1(a - x_e) = k_{-1}x_e^2$$

that is

$$k_1 a - k_1 x_e - k_{-1}x_e^2 = 0 \tag{11.1}$$

Let the deviation from equilibrium be Δx where $\Delta x = x - x_e$. Therefore

$$\frac{d(\Delta x)}{dt} = \frac{dx}{dt} = k_1 a - k_1 x - k_{-1}x^2 \tag{11.2}$$

$$= k_1 a - k(\Delta x + x_e) - k_{-1}(\Delta x + x_e)^2$$

Combining equations 11.1 and 11.2 and neglecting squared terms involving Δx gives

$$\frac{d(\Delta x)}{dt} = -(k_1 + 2k_{-1}x_e)\Delta x \tag{11.3}$$

Integrating equation 11.3 gives

$$\ln \Delta x = -(k_1 + 2k_{-1}x_e)t + \text{constant}$$

When $t = 0$, $\Delta x = (\Delta x)_0$, so that constant $= \ln(\Delta x)_0$. Therefore

$$\ln \frac{(\Delta x)_0}{\Delta x} = +(k_1 + 2k_{-1}x_e)t$$

The *relaxation time* is defined as the time needed for a system to traverse a fraction $1/e$ of its path to equilibrium. Therefore

$$\tau = \frac{1}{k_1 + 2k_{-1}x_e}$$

A measurement of the relaxation time combined with the determination of the equilibrium constant k_1/k_{-1} enables the individual rate constants k_1 and k_{-1} to be calculated. Using this method the rate constants for the reaction

$$HA + H_2O \rightleftharpoons H_3O^+ + A^-$$

$k_1 = 7.8 \times 10^5 \, s^{-1}$ and $k_2 = 4.5 \times 10^{10} \, dm^3 \, mol^{-1} s^{-1}$, were measured.

The essential requirement for the measurement of a relaxation time is a method that gives a sharp disturbance followed by a fast *in situ* analysis. Three techniques are commonly used.

(i) Temperature jump

A sudden electric current is passed through the sample, and the resultant ions collide with solvent molecules and thereby heat the solution. If a microwave pulse is used, rotational energy transfer from polar solvent molecules heats the solution. An infra-red laser pulse causes the transfer of vibrational energy from the solvent molecules to the solution. By these techniques it is possible to induce a temperature rise of up to $10°C$ in a microsecond. A number of enzyme catalyses has been studied in this way.

(ii) Pressure jump

This is effective for reactions with relatively high volume changes. The most convenient method of inducing a pressure jump is by a shock wave or a release of hydrostatic pressure from a burst diaphragm.

(iii) Electric field pulse

When the solution of a weak electrolyte is subjected to a very large electric field (say $10^5 \, V \, cm^{-1}$), the equilibrium is disturbed and the acid dissociation constant increases. This method has been used to measure rate constants of many protonation and deprotonation reactions, all of which are about $10^{10} \, dm^3 \, mol^{-1} s^{-1}$. Such reactions are diffusion controlled.

Further reading

Reference

1. B. S. Yamanashi and A. W. Nowak, *J. chem. Ed.,* 45 (1968), 705.
2. J. N. Bradley and G. B. Kistiakowsky, *J. chem. Phys.,* 35 (1961), 256.

Reviews

D. N. Hague. Experimental methods for the study of fast reactions. *Comprehensive Chemical Kinetics,* 1 (1969), 112.
A. Weller. Fast reactions of excited molecules. *Progr. Reaction Kinetics,* 1 (1963), 187.

G. Dixon-Lewis and A. Williams. Methods of studying chemical kinetics in flames. *Q. Rev. chem. Soc.*, **17** (1961), 243.

R. G. W. Norrish and B. A. Thrush. Flash photolysis and kinetic spectroscopy. *Q. Rev. chem. Soc.*, **10** (1956), 149.

G. Porter. Flash photolysis and some of its applications. *Science*, **160** (1968), 1299.

L. M. Dorfman and M. S. Matheson. Pulse radiolysis. *Progr. Reaction Kinetics*, **3** (1965), 237.

F. W. Willetts. The evolution of flash photolysis and laser photolysis techniques. *Progr. Reaction Kinetics*, **6** (1971), 1.

H. O. Pritchard. Shock waves. *Q. Rev. chem. Soc.*, **14** (1960), 46.

A. Blythe, M. A. D. Fluendy and K. P. Lawley. Molecular beam scattering. *Q. Rev. chem. Soc.*, **20** (1966), 465.

E. F. Greene and J. Ross. Molecular beams and a chemical reaction. *Science*, **159** (1968), 587.

Books

E. Caldin, *Fast Reactions in Solution*, Blackwell, Oxford (1964).

S. Claesson, *Fast Reactions and Primary Processes in Chemical Kinetics*, Interscience, New York (1967).

D. N. Hague, *Fast Reactions*, Wiley, London (1971).

APPENDIX

SI Units

Throughout, a single system of units known as SI (abbreviation for *Système International d'Unités*) has been used. SI is based on the following seven independent physical quantities.

Physical quantity	Symbol	Basic SI unit	Unit symbol
length	l, etc.	metre	m
mass	m	kilogramme	kg
time	t	second	s
electric current	I	ampere	A
thermodynamic temperature	T	Kelvin	K
amount of substance	n	mole	mol
luminous intensity	I_v	candela	cd

Other physical quantities can be expressed in SI units which are derived from the above by appropriate multiplication, division, integration and/or differentiation without the introduction of any numerical factors (including powers of ten). A selection of those used frequently throughout are given below.

Physical quantity	Symbol	Name of unit	Unit symbol
force	F	newton	$N = kg\, m\, s^{-2}$
pressure	p	–	$N\, m^{-2} = kg\, m^{-1}\, s^{-2}$
energy (all forms)	H, U, E	joule	$J = N\, m = kg\, m^2\, s^{-2}$
entropy	S	–	$J\, K^{-1}$
rate	$\dfrac{dc}{dt}$	–	$mol\, m^{-3}\, s^{-1}$
rate constant of $(n + 1)$th order reaction	k_r	–	$m^{3n}\, mol^{-n}\, s^{-1}$
activation energy	$E^{\ddagger}$	–	$J\, mol^{-1}$
collision rate	Z	–	$m^{-3}\, s^{-1}$
quantum yield	ϕ	–	dimensionless
frequency	ν	hertz	s^{-1}

Non-basic SI units

Decimal fractions and multiples of basic SI units can be denoted by means of prefixes. In many cases it is convenient insofar as it avoids the use of inconveniently large and small numerical values. In particular, concentrations are normally expressed in $mol\,dm^{-3}$ units. For example, a concentration of $0.1\,mol\,dm^{-3}$ is preferable to $100\,mol\,m^{-3}$ because (a) $1\,m^3$ of solution is in considerable excess of the scale of a typical kinetics experiment, (b) it is numerically equal to the obsolete term, *molar concentration*, and (c) it is numerically similar to the molality expressed in $mol\,kg^{-1}$.

However, it is important to realise that when performing numerical calculations, the data should be converted to unprefixed SI units. This eliminates the need for conversion factors and reduces the risk of miscalculation.

A rate of reaction is therefore usually expressed in $mol\,dm^{-3}\,s^{-1}$ and a second-order rate constant in $dm^3\,mol^{-1}\,s^{-1}$.

Non-SI units

With the exception of the atmosphere, which is retained in view of its role as a standard state, most non-SI units are unnecessary. A number, such as minute, hour, degree Celsius will continue in everyday use, and have been retained in the text.

In many cases it may be convenient to express experimental data in non-SI units in view of the method of measurement. However, non-SI units should in general be avoided when calculating physical quantities from the experimental data. For example, if a mercury manometer is used to measure the pressure variation of a reaction at constant temperature and volume over a period of several minutes it is convenient and permissible to tabulate pressures in mm of Hg or torr, with the corresponding reaction times in min. If the reaction is first order, a conversion to SI units is not necessary (see example 3.7), and the rate constant has units of reciprocal time. However, if the reaction is second order, conversion will be necessary, since the rate constant should be expressed in $N^{-1}\,m^2\,s^{-1}$ but not in $torr^{-1}\,min^{-1}$.

Methods of expression

Tables

The value of a physical quantity is expressed as the product of a pure number and a unit. For example

$$k_r = 1.41 \times 10^{-3}\,dm^3\,mol^{-1}\,s^{-1}$$

This can rearrange to

$$k_r/dm^3\,mol^{-1}\,s^{-1} = 1.41 \times 10^{-3}$$

or

$$10^3 k_r/\text{dm}^3\,\text{mol}^{-1}\,\text{s}^{-1} = 1.41$$

To avoid repetition of a unit symbol, it is normal to tabulate data in the form of pure numbers. It follows that the column headings should be dimensionless; for example, $k_r/\text{dm}^3\,\text{mol}^{-1}\,\text{s}^{-1}$, not $k_r(\text{dm}^3\,\text{mol}^{-1}\,\text{s}^{-1})$, since the latter implies the rate constant multiplied by $\text{dm}^3\,\text{mol}^{-1}\,\text{s}^{-1}$.

Since only pure numbers can be converted into the corresponding logarithms, it follows that tables of logarithmetic expressions should also be dimensionless.

An example of the notation is taken from example 4.1.

$10^3 k_r/\text{dm}^3\,\text{mol}^{-1}\,\text{s}^{-1}$	T/K	$10^3\text{K}/T$	$\log_{10}(k_r/\text{dm}^3\,\text{mol}^{-1}\,\text{s}^{-1})$
0.100	293	3.413	4.000
0.355	303	3.300	4.525
1.41	313	3.195	3.015
3.06	323	3.096	3.485
8.13	333	3.003	3.910
21.1	343	2.915	2.325
50.1	353	2.833	2.700

The above considerations apply to the labelling of graphs used.

Table of physical constants

Constant	Symbol	Value	Logarithm
Velocity of light in a vacuum	c	$2.9979 \times 10^8 \, \text{m s}^{-1}$	8.4768
Permeability of a vacuum	μ_0	$4\pi \times 10^{-7} \, \text{H m}^{-1}$	$\bar{6}.0992$
Permittivity of a vacuum	ϵ_0	$8.8542 \times 10^{-12} \, \text{F m}^{-1}$	$\overline{12}.9471$
Rest mass of electron	m_e	$9.1091 \times 10^{-31} \, \text{kg}$	$\overline{31}.9595$
Elementary charge	$e = F/N_A$	$1.6021 \times 10^{-19} \, \text{C}$	$\overline{19}.2047$
Boltzmann constant	$k = R/N_A$	$1.3805 \times 10^{-23} \, \text{J K}^{-1}$	$\overline{23}.1400$
Planck constant	h	$6.6256 \times 10^{-34} \, \text{J s}$	$\overline{34}.8212$
Avogadro constant	N_A	$6.0225 \times 10^{23} \, \text{mol}^{-1}$	23.7798
Faraday constant	$F = N_A e$	$9.6487 \times 10^4 \, \text{C mol}^{-1}$	4.9845
Gas constant	$R = N_A k$	$8.314 \, \text{J K}^{-1} \text{mol}^{-1}$	0.9198
Molar volume of ideal gas at $0\,^\circ\text{C}$ and 1 atm		$2.2414 \times 10^{-2} \, \text{m}^3 \text{mol}^{-1}$	$\bar{2}.3505$
Standard gravitational acceleration	g	$9.8066 \, \text{m s}^{-2}$	0.9915

Conversion factors

	Logarithm
$1 \, \text{atm} = 1.01325 \times 10^5 \, \text{N m}^{-2}$	5.0057
$1 \, \text{torr} = \frac{1}{760} \, \text{atm} = 133.32 \, \text{N m}^{-2}$	2.1249
$1 \, \text{eV} = 1.6021 \times 10^{-19} \, \text{J}$	$\overline{19}.2047$
$0\,^\circ\text{C} = 273.15 \, \text{K}$	2.4364
$1 \, \text{kcal mol}^{-1} = 4.182 \, \text{kJ mol}^{-1}$	0.6214

ANSWERS TO PROBLEMS

Chapter 2 (page 24)

1. $k_r = 2.22 \times 10^{-8} \, (N\,m^{-2})^{-1} s^{-1}; \dfrac{d[I_2]}{dt} = 0.722 \, N\,m^{-2}\,s^{-1}$
2. (a) Second order with respect to NO and first order with respect to H_2
   (b) $k_r = 145.5 \, dm^6 \, mol^{-2} \, s^{-1}$
3. $k_r = 4.37 \times 10^{-2} \, min^{-1}$
4. $k_r = 3.55 \times 10^{-3} \, min^{-1}$
5. Reaction is second order; $k_r = 2.13 \times 10^{-3} \, dm^3 \, mol^{-1} s^{-1}$
6. $k_r = 2.48 \times 10^{-2} \, dm^3 \, mol^{-1} \, min^{-1}$
7. $k_r = 0.667 \, dm^3 \, mol^{-1} s^{-1}$
8. $k_r = 9.2 \times 10^{-3} \, dm^3 \, mol^{-1} \, min^{-1}$
9. $k_r = 0.33 \, dm^3 \, mol^{-1} \, min^{-1}$
10. Zero order

Chapter 3 (page 44)

1. Excess iodide, therefore first-order reaction; $k_r = 1.23 \times 10^{-3} \, s^{-1}$
2. $k_r = 8.12 \times 10^{-2} \, dm^3 \, mol^{-1} s^{-1}$
3. $k_r = 6.74 \times 10^{-2} \, min^{-1}$
4. $k_1 + k_{-1} = 8.10 \times 10^{-5} \, s^{-1}$
5. $k_1 + k_{-1} = 1.46 \times 10^{-2} \, min^{-1}$
6. $k_r = 5.07 \times 10^{-3} \, min^{-1}$
7. $k_r = 1.78 \times 10^{-2} \, s^{-1}$
8. $k_r = 0.134 \, min^{-1}$
9. Plot of $\log_{10}(2p_0 - p)$ against t is linear, indicating a first-order reaction; $k_r = 1.43 \times 10^{-2} \, min^{-1}$
10. Plot of $\log_{10}(3p_0 - p)$ against t is linear, indicating a first-order reaction; $k_r = 8.63 \times 10^{-2} \, min^{-1}$

Chapter 4 (page 57)

1. $E^{\ddagger} = 304 \, kJ \, mol^{-1}$
2. $E^{\ddagger} = 196 \, kJ \, mol^{-1}$
3. $E^{\ddagger} = 1041 \, kJ \, mol^{-1}$
4. $E^{\ddagger} = 146 \, kJ \, mol^{-1}$
5. $E^{\ddagger} = 52.9 \, kJ \, mol^{-1}$

6. 36.2 per cent of the 20 per cent solution
7. 617 K

Chapter 5 (page 70)
1. $k_r = 8.4 \times 10^{-6} \, m^3 \, mol^{-1} \, s^{-1}$
2. $\Delta S^{\ddagger} = -187 \, J \, K^{-1} \, mol^{-1}$
3. $\Delta S^{\ddagger} = 46.6 \, J \, K^{-1} \, mol^{-1}$
4. $\Delta S^{\ddagger} = -72.0 \, J \, K^{-1} \, mol^{-1}$

Chapter 6 (page 79)
1. $k_{\infty} = 8.48 \times 10^{-4} \, s^{-1}; k_1 = 0.284 \, dm^3 \, mol^{-1} \, s^{-1}$
2. $k_1 = 9.9 \times 10^{-12} \, dm^3 \, mol^{-1} \, s^{-1}$
3. 1.07×10^{-7}

Chapter 7 (page 96)
1. $\dfrac{d[H_2]}{dt} = k_3 \left(\dfrac{k_1}{k_5}\right)^{1/2} [C_2H_6]^{1/2}$

3. $\dfrac{-d[O_3]}{dt} = \dfrac{2k_1 k_2 [O_3]^2}{k_{-1}[O_2] + k_2}$. Therefore, second order with respect to ozone and the rate is inhibited by oxygen

4. $\dfrac{-d[ROOR]}{dt} = k_1[ROOR] + k'[ROOR]^{3/2}$, where $k' = k_3 \left(\dfrac{k_1}{k_4}\right)^{1/2}$

5. $x = 1.0$ (assuming constant initiator concentration); $y = 0.5$

$$
\begin{array}{ll}
\text{Mechanism} & \quad I \to 2R \quad \text{(initiation)} \\
& \left.\begin{array}{l} R + M \to R_1 \\ R_1 + M \to R_2 \\ \cdots\cdots\cdots\cdots \\ R_{n-1} + M \to R_n \end{array}\right\} \text{(propagation)} \\
& \left.\begin{array}{l} R_n + R_m \to P_{n+m} \\ R_n + R_m \to P_n + P_m \end{array}\right\} \text{(termination)}
\end{array}
$$

Chapter 8 (page 111)
1. $z_A z_B = 2.0; k_r = 8.41 \times 10^{-2} \, dm^3 \, mol^{-1} \, min^{-1}$
2. (a) $k_0 = 1.89 \, dm^3 \, mol^{-1} \, s^{-1}$. (b) $k_r = 1.28 \, dm^3 \, mol^{-1} \, s^{-1}$
3. $\Delta V^{\ddagger} = -13.1 \times 10^{-6} \, m^3 \, mol^{-1}$

Chapter 9 (page 131)
1. $k_r = 4.1 \times 10^{-2} \, dm^3 \, mol^{-1} \, min^{-1}$
2. $k_r = 8.0 \times 10^{-3} \, dm^3 \, mol^{-1} \, s^{-1}$
3. (a) Zero order. (b) First order at low pressures, zero order at moderate pressures. (c) At low pressures the rate is proportional to the fraction of surface covered. At moderate pressures, the rate is independent of pressure because NH_3 molecules are adsorbed on all the available sites.
4. $K = 4.8 \times 10^{-4} \, mol \, dm^{-3}; v_{max} = 1.22 \, \mu mol \, s^{-1}$

Chapter 10 (page 145)
1. $\phi = 0.094$
2. 2.55×10^{21} molecules
3. 1.22×10^{-8} mol
4. Steady-state treatment gives $[I]^2 = \dfrac{I_{abs}}{k_2[I_2] + k_3[H_2] + k_4[H_2]}$

5. $\dfrac{d[HBr]}{dt} = \dfrac{k_2(I_{abs}/k_{-1})^{1/2}[H_2][Br_2]^{1/2}}{1 + \dfrac{k_3[HBr]}{k_{-2}[Br_2]}}$

INDEX